PREFACE AND BACKGROUND
Tim Bostock[1]

This book represents the final output of a policy study that was initiated in 2002 under the auspices of an agreement between the Support unit of International Fisheries and Aquatic Research (SIFAR)[2] and The World Bank[3]. In the past, fisheries management tends to have been characterised by failure rather than by success. Conventional responses by well-intentioned development agencies have therefore focused on identifying the causes of this failure and attempting to rectify them. The idea underlying the study and this book was to explore an alternative route to the same objective, i.e. improved fisheries management, by attempting to identify success in fisheries management and the factors that have led to it.

The study was informed by a series of seven case studies implemented in selected fisheries from around the globe, each of which was recognised as demonstrating some facets of success. These included: Pacific halibut fishery; fisheries sector of Mauritania; co-managed and community-based fisheries in Shetland; traditional community-based management in Andhra Pradesh, India; Senegalese artisanal fisheries of Kayar; Namibian hake fishery; and the Australian northern prawn fishery. Core principles from each specific experience were distilled through a consultation process, and these then informed generic conclusions on 'management good practice'.

Each case study illustrates one or more of the several possible dimensions of success. In this context, it is instructive to examine what we mean by 'success', as the measure of this will differ according to the specific management and policy objectives in any particular situation. 'Success' has at least three dimensions[4]: biological (meeting conservation objectives); economic (meeting wealth and efficiency objectives) and social (meeting the equity objectives, both in terms of distribution and access). Successful fisheries thus require institutional capacity both to define an appropriate balance of these parameters within management objectives, and to implement and adapt these responsively over time. Moreover, if fisheries are to

---

1 At the time of this study, Tim Bostock was coordinator of the SIFAR programme. He is currently senior fisheries adviser at the Department for International Development, UK.
2 The SIFAR programme was based at FAO, Rome from 1998 to 2004 and was engaged in facilitating a range of research-based activities focused on addressing the goal of sustainable, equitable and optimal fisheries management.
3 I would like to take this opportunity to thank the World Bank, in particular Cees de Haan, for the support and interest shown in this project, the output from which was used to inform the World Bank's 2004 Approach paper on fisheries "Saving Fish and Fishers" Report No. 29090-GLB.
4 Another dimension might be considered as 'political success'.

contribute successfully to growth (and particularly pro-poor growth in developing countries), such capacity is of central importance.

Success also has several causes which are well illustrated by the case studies. While some of these relate to the selection of management instruments, others depend more fundamentally on the prevailing governance environment - the effectiveness of existing institutional arrangements and policy processes. It is evident from this, that an understanding of success predetermines an understanding of failure. In this regard, a key finding of the study was that weak or dysfunctional governance systems result in fishers operating under perverse, unsustainable incentive structures that engender a 'race for fish'. Overcoming this requires an institutional and policy environment that instead creates incentives to conserve fish and optimise wealth generation.

Although some of the tools that can power such a change are already available and are well-illustrated in the case studies, many of these (e.g. permanent rights, fiscal reforms and decentralization) require an about-turn in political thinking. Not only do they involve painful decisions (capacity reduction, institutional reform, ...), but they also represent radical departures from 'business-as-usual' approaches to conventional fisheries management. To be successfully applied, political will and commitment are essential. The book concludes by underscoring the major challenge for those involved in the development and promotion of more effective fisheries management: this is not to do the same things better, but to do them differently.

Overview

Stephen Cunningham[1]

Bibliographic searches of the fisheries literature for evidence of, or even reference to, success in fisheries management produce very little. There is an overwhelming concentration on bad news with innumerable studies demonstrating disaster, failure and human-error. This result does not come as a great surprise, partly because it is widely acknowledged that fisheries management has been characterised more by its failures than its successes and partly because even where authors identify some element of success in management arrangements there is an irresistible urge to offer advice to improve things. The usual approach is to attempt to identify what is wrong with fisheries management and offer suggestions as to how to put it right.

This study attempts to take a different route by identifying what is right with fisheries management in particular instances and offering suggestions as to how to build on such success. It study is based around a series of case studies supported by a literature review.

This approach requires an examination of what is meant by 'success' in fisheries management. Where such success has been achieved, there is a need to identify the factors responsible for it, distinguishing between those which are location-specific and those which offer general lessons. Most importantly, there is a need to consider how success can be achieved on a sustainable basis.

THE NATURE OF SUCCESS
The literature review highlights some of the characteristics of success in fisheries management which are summarised below:

Multi-dimensional
A key issue is that success in fisheries management is multi-dimensional. The principal dimensions appear to be social, economic and biological.

Social success requires that fish resources contribute to social welfare in a manner considered to be equitable. The precise nature of this contribution will vary from fishery to fishery, from place to place and possibly from time to time. Through the political process, choices will have to be made. For instance, in some places it may be preferred to use fish resources to provide livelihoods to resource-dependent communities; in others the choice may be to use the wealth of the resource more widely.

---

1 Institut du Développement Durable et des Ressources Aquatiques (IDDRA), Montpellier, France and Portsmouth, UK.

The range of choice available will depend on achieving economic success, i.e. fisheries which are operated in an efficient manner, avoiding problems of overcapacity. A key issue in economics is the resource rent that the fishery is capable of generating. Success requires that this rent is either capitalised (into the price of a right) or extracted (as a royalty). Leaving the rent in the fishery leads to its dissipation through excessive levels of exploitation.

Another important issue is the relationship between resource rent and social welfare. The assumption, implicit or otherwise, is often made that social welfare will be maximised if resource rent is maximised (economic efficiency maximisation). Since the latter will be achieved by the unrestricted transferability of permanent fishing rights, these kinds of rights are often suggested (at least by economists). However, social success may require that some trade-offs be made between equity and efficiency, which could require for instance some restrictions on transferability to protect vulnerable groups.

Underpinning social and economic success will be biological success. Such success requires that overexploited fish stocks be rebuilt because in many cases higher stock levels will enable greater catches to be taken on a sustainable basis in the long run with less variability.

Successful fisheries management must strike a balance between these different dimensions. In order to do this, managers must ensure that appropriate information is available on each dimension. They must also ensure that there is a correct identification of objectives and constraints.

The multi-dimensional nature of fisheries management may also be one explanation of why there appears to be so little success at present. Since failure is generally, if not always, the obverse of success, it too is multi-dimensional. Suppose that in a particular situation it is possible to identify 10 dimensions to the problem. Suppose also that 9 may be characterised as successful. The overall situation may, nonetheless, be considered a failure because much more stress is likely to be laid on the failure in the 10th dimension than on the 9 successful ones. Interest groups are likely to emerge to lay emphasis on the failed dimension.

One way out of this dilemma would be to agree on appropriate dimensions and on success criteria for each dimension. Such an approach would enable a more nuanced evaluation of fisheries management systems than often appears to be the case at present. Rather than simply categorising systems as success or failure (where the latter category includes many more outcomes than the former), it would be possible to provide a qualitative ranking system along the lines of "failure", "largely unsuccessful", "moderately successful", "largely successful", "success". It would also be instructive to give some idea of movement between categories since it clearly makes a difference whether things are improving or not.

The development of the dimensions and criteria will raise the issue that not all dimensions are equal. Some will be objectives whereas others are constraints - the most obvious being the resource constraint. The difficulty of the conflict between objectives would also have to be addressed, as would the possibility of trading-off between dimensions. To what extent, for instance, is it acceptable to increase exploitation beyond the level that would be dictated by purely biological factors if such increased exploitation enables important socio-economic objectives (such as employment) to be achieved. Criticisms levelled at fishery management systems often give the impression that trade-offs are unacceptable.

In order to improve the performance of such an indicator, a number of other things would have to be known - for instance, has the fishery management authority deliberately chosen to operate at a particular level because it can thereby achieve other socio-economic goals? Is the situation sustainable? Is the situation improving? Whilst it is clearly true that there are many cases where fisheries management has failed, there are also many cases where accusations of failure have been used to disguise a debate about management priorities.

One practical suggestion that emerges from this discussion would be to undertake an evaluation of fishery management systems using a wider set of criteria than simply the state of the fish stocks (important though this latter clearly is). The impression gained as to whether management is broadly successful or unsuccessful, and broadly improving or deteriorating may be rather different from the current pervasive impression of failure.

Relationship to objectives

A standard way of evaluating success is to compare outcomes to objectives. The drawback with this approach is that if objectives are mis-specified then success in achieving them may still not equate to successful fisheries management. This is a difficult area because the impression may be given that policy analysts are attempting to impose objectives on decision-makers. This is not the intent. Rather, it is to ensure that policy-makers analyse the full impact of attempting to achieve particular goals on all dimensions of the system. The Pacific halibut case clearly demonstrates for instance that over-emphasis on biological goals may lead to success in that dimension but at the expense of failure in other dimensions, particularly the economic one, leading ultimately to the failure of the management system. Similar problems arise in many cases where targets are production-based.

The litmus test for successful fisheries management is perhaps best assessed therefore in terms of the sustainability of the management arrangements themselves, since sustainable arrangements are likely to be those which include the various dimensions of sustainability.

In the same vein, there is a need to distinguish between cause and effect, or cause and symptom, in order to design effective fishery management systems. The failure criterion to which attention is most frequently drawn is the fish stock itself. Yet (except for very extreme cases where fishing to extinction is profitable) overfishing the fish stock is in the interest of no one. Such overfishing arises because of a failure somewhere else, in some other dimension of the system, most notably in the area of incentives. Correcting such incentives may be expected to correct the overfishing problem. Policy measures, however, are frequently designed and recommended as if the failure to conserve fish stocks is a stand-alone problem to which measures should be applied directly. Throughout much of Northwest Africa there is currently an obsession with "biological rest periods", which are in fact closed seasons by another name. The main objective of these closures is to lower fishing mortality but since they do not deal with the fundamental failure, the easily-predicted outcome is observed that, in order to achieve a given impact, they have to become increasingly long.

Dynamic nature of success
To be successful, a management system must be anticipatory in lots of ways. One important element is that managers must anticipate that the definition of the problem itself will evolve over time.

Successful fisheries management used to turn around one fish stock and the fishers exploiting it. Over the past few years, it has become commonplace to chastise management for not taking an "ecosystem-based" approach to the problem. Managers who felt that they were successfully managing their fisheries now find that they are failing in this new dimension, or newly prioritised dimension. In fact, because the structure of the management problem is arbitrary, it is always possible to re-define it so as to render the current situation one of failure.

Evaluating the success of management is also complicated by the fact that the state of the system may change for reasons that have nothing to do with management. This problems complicates the issue in two ways. First, managers may be blamed (take credit) for negative (positive) effects that are unrelated to their actions. Second, when things go wrong they may look around for external factors to blame. Analyses of collapsed fisheries almost always list a host of external factors (temperature changes, salinity changes, climate change, El Niño etc), whereas the one common factor tends to be overexploitation.

Wealth issues
Resource rent is clearly a key element in successful fisheries management. A first requirement is for management instruments to be implemented which allow such rent to be generated.

It is also very important to address the distributional question at the outset because experience strongly suggests that inequitable systems (which largely means in terms of benefit sharing, although may also mean access arrangements) may not be sustainable. Great care needs to be taken concerning how the instruments are implemented because it may be very difficult and/or expensive to change the implementation. For instance, in Iceland ITQs were given to fishers. Once it became clear how much wealth had been given away calls were made, and continue to be made, for such rights to be recovered by the State on behalf of the general public. The problem is that because rights were granted in perpetuity, expected rents have been capitalised into their price. Moreover because some rights have been traded, rights owners now comprise a mix of those who have bought rights and those who received them for free. An equitable solution to the question of recovering rights does not seem to exist, or at least is not obvious. Countries that are considering rights-based solutions to management need to address at the outset the issue of equity and how the rents are to be distributed.

It could be argued that fish resources are the property of the nation and are managed by Government on behalf of the nation. Statements such as this are often found in fisheries legislation. On a strict reading, such statements would imply that all rents should be extracted by the Government on behalf of the nation as resource owner. However, there are reasons for thinking that a fair share of rents between the nation and fishers is likely to be in the general interest.

First, as a practical matter, taxing resource rent at 100% is extremely difficult, if not impossible. The inherent variability in fishing means that the taxation system would have to have a degree of flexibility which rarely if ever exists in such systems.

Second, experience around the world clearly shows that very high levels of tax (whatever the tax base) encourage the private sector to adopt tax avoidance behaviour (legal or not) which is very expensive to control. As a result, net tax revenue is often higher at lower tax rates.

Third, and the most important argument from a fishery management viewpoint, leaving the private sector with a share of the resource rent gives fishers a stake in the well-being of the fish stock and in the economic improvement of its exploitation. If a share is to be left to the private sector on a sustainable basis, a successful management system must find an instrument enabling the value of the rent to be revealed. In many countries the chosen instrument is the ITQ. In this case, fishers should have an incentive towards sustainable exploitation so as to maintain and increase the value of their use rights. This stake in the future of the resource may also help with enforcement of management measures since non-compliance will impact on the value of ITQs.

In conclusion, successful management needs to address the wealth issue very early on. The fact that many fisheries are in poor economic condition when management is implemented should not blind the authorities to this need. Correction after the event is often not an option.

Lessons from the case studies

The case studies undertaken in the context of this work illustrate different features of success. Given their wide variety, the first point that they underline is that there is no unique recipe for success. It is a question of designing a system that is appropriate to the particular circumstances of each fishery. The case studies are briefly as follows.

Pacific Halibut

The management of the Pacific halibut (*Hippoglossus stenolepis*) fishery has received considerable attention over the years. The Pacific halibut fishery is sometimes regarded as a model fishery in terms of biological conservation, but less successful in terms of economic objectives such as cost effective harvesting. This is largely because until recently management focussed on one dimension of the problem: stock conservation. But this approach was ultimately unsuccessful because of distortions introduced elsewhere, notably dramatic reductions in season length leading to other problems (such as unsafe fishing leading to loss of life every year and reductions in fish quality and consumer satisfaction). As fishing days were reduced, input stuffing took place to 'make up' for lost fishing time.

By the mid 1990s, it was realised that to achieve both biological and economic objectives, without the help of rights-based fishing, would be practically impossible. In 1995, the official season had been reduced to two days. What had once been an annual fishery had been reduced to a classic 'derby' fishery with crews working non-stop for 48 hours without due regard for dangerous sea conditions and overloaded vessels. The season's tally, duly reported in the media would include vessels and lives lost due to the dangerous operating conditions and the need to survive economically in the regulatory regime forced on the vessels. With limited time, each minute became precious; crews would cut entangled long-lines, rather than disentangle them, and the lines would drift, continuing to 'ghost fish' for whatever species swallowed the bait and hooks. Gentler handling and release of undersized halibut became an unaffordable use of scarce crew time so such fish were torn apart and discarded with attendant low survival rate.

The marketing outcomes of reduced seasons were also devastating. Total annual catch was landed at the end of a single fleet trip. Halibut is far superior as a fresh product but, due to the regulatory approach, the product mix was tilted almost entirely toward inferior frozen forms which had to be stored at high cost to serve in future markets. Under a different regulatory framework, a large fraction of the same harvests would have been left in the ocean until needed to meet the demand for high quality fresh fish.

Recognising the inherent problems with the 'race to fish' Canadian fishermen asked for individual quotas which were implemented in 1991 and Individual Fishing Quotas (IFQs) were introduced in Alaska in 1995 to replicate the Canadian success at extending the fishing period and making fishing operations much safer.

The introduction of rights-based fishing has had numerous positive effects. Stock recovery has been noted, landings have increased, prices on markets have improved and stabilised, the fishing season has lengthened with knock-on social effects in terms of safety at sea. The freedom to fish more selectively has led both to a reduction in discards and in ghost fishing.

Mauritania

The Mauritanian case study is interesting because it illustrates that fiscal instruments can be used to manage fisheries successfully. There appear to be very few cases around the world where this has been done, although with the development of "green taxes" associated with the environmental lobby there may be more interest in this approach in the future.

Mauritania's system was very simple but appropriate to the circumstances of its fishery (almost totally for export from very few landing points). A state monopsony was set up through which fish was channelled prior to export. This monopsony went through various incarnations but the most successful appears to have been when it bought fish from domestic fishers and then sold it into export markets.

Although the system was not developed to its full potential, during the period to 1995, differential tax rates were used as a means both to extract resource rents and implement fisheries policy. Taxes were higher on products that were frozen at sea and exported so as to encourage development of processing in Mauritania, and were higher on cephalopods than other fish to discourage excessive targeting of cephalopods.

This system was replaced in 1995 by a licensing system that was supposed to be fiscally neutral. As the case study makes clear this objective has not been achieved and the licensing system has been far less successful than the system it replaced.

The tax-based system was successful in a number of ways. By extracting resource rent, around 20% of total government revenue came from taxes through the monopsony system in the mid-1980s. Over time, fishing agreements became of more importance, but even in 1995 (the year the system was abandoned) around 15% of government revenue still came from this source. The licensing system has not been successful and in 2000 only just over 2% of government revenue came from the domestic fishing industry, fishing agreements having increased to 16%. The result is that the government has become increasingly dependent on such agreements. The monopsony system enabled Mauritania to ensure that the fish resource contributed appropriately to foreign reserves and favoured the development of the artisanal fishery mainly because it provided a ready market for artisanal produce. The monopsony worked with artisanal fishers to ensure that their product quality met the export-market standards. It also provided countervailing power to the large buyers on the export market and in particular prevented fruitless competition between Mauritanian sellers (since they were price takers).

Shetlands

The Shetland case study illustrates different points. The main theme is the way in which stakeholder participation coupled with strong local leadership can lead to the emergence of successful management systems.

The study offers three different examples, each with different lessons. The first case concerns the small industrial fishery for sand eels where commercial fishermen and environmentalists have formed a partnership. The sand eel fishery was closed in 1991 following a rapid decline in catches during the 1980s. The decline was associated with increased chick mortality in a number of Shetland seabird colonies and a fierce debate ensued between environmental organisations that argued that chick mortality was a direct consequence of the industrial fishery, and the fishing industry, which argued that the chick mortalities and the decline in fishery were both related to a decline in stock. After extensive negotiations between environmental groups and the local fishermen, the fishery was eventually reopened in 1995 with reduced quotas, vessels limits and closed seasons to protect the birds. Moreover, management of the fishery was moved to the local fishing industry.

In the second case, the way the Shetland shellfish industry was managed was radically changed when industry representatives successfully applied to be able to manage the stocks locally through a regulatory order. The third case considers how the management of demersal and pelagic fisheries was altered to allow the Shetlands to manage the haddock quota on behalf of the community. This experiment, in so far as the Shetland fishing industry was concerned, was successful. As a result there was a rapid move towards adopting sectoral quotas throughout the fishing industry at large.

India
This case shows how two communities with different management systems have been able to respond to pressures and changes. The two communities are Uppada and Boddu Chinna Venkataya Palem (BCV Palem). The first is a sea fishing community which has changed considerably over time: technological advances have influenced the types of boats used and the variety of fishing gear, and coastal erosion has seen the village move several times. The increasing globalisation of the fishing industry, particularly the advent of shrimp fishing, has impacted upon Uppada (e.g. affecting catch prices). Alternative communal resources are limited in Uppada. The second community is riverine with a far smaller population. Because of the nature of creek fishing in shallow waters, technology has had little impact here and the levels of rent generated by the fishery are much smaller. However, the community does have access to considerable sources of alternative communal resources (e.g. extensive tracts of mangrove).

Fisheries in both communities are managed through TCBMS, which are rarely sector-specific to fishing, but provide a framework for managing livelihoods and the community as a whole. As such they generally encompass the socio-cultural, religious, political, administrative and economic aspect of the community organisation and within this context, fishing activity is also managed.

The specific nature of TCBMS varies between fisheries. In BCV Palem, where fishing activities are carried out by a number of fishing systems confined to the creeks and the backwaters, the traditional systems of management and control related to fisheries and fishing are elaborate and have an important economic function. In Uppada, on the other hand, where there is often considerable competition for space for beach seining which often requires large groups of people, there is a greater emphasis on social issues and relationships.

In both villages, the Panchayats (upon which the TCBMS is based) serve an important welfare and equity function. The monies generated from renting out or otherwise making use of the village commons are shared amongst all members of the community. The amount generated in BCV Palem is much larger than in Uppada but any surplus generated from leasing/renting out of these lands is distributed to members of the Panchayat. Considering the benefits, which cover important aspects of their life and livelihoods, people have little difficulty in willingly and implicitly followed the caste code. Another important source of income to the Panchayat is the revenue generated from fines. In BCV Palem, the income from fines takes care of the elders' time, because they share the amount equally amongst themselves. In Uppada, this may have been the most important source of income to the Panchayat, which lacks any other productive sources.

Though the villages were largely homogeneous in social and economic terms, they were by no means peaceful, and there were always disputes - most of them petty. The foremost activity of the village Panchayat was settling the village disputes related to fishing: boundary disputes, disputes concerning transactions between boat owners and their crew and conflicts concerning outsiders. The Panchayat thus acted as a link between the community and the external world, taking the community grievances to the government functionaries, who generally resided elsewhere, and acting as the first point of contact for anyone visiting the village from outside.

Kayar, Senegal

Senegalese fisheries are characterised by over-exploitation of demersal coastal resources and over capacity in the fishery. This has translated into reduced catches and a rise in fisheries conflicts between the different sections of the fleet competing for dwindling stocks. A reaction to this situation in the village of Kayar is instructive.

Kayar is on the northern coast of Senegal about 58 km from the capital Dakar. Like other villages, it was badly hit by the sharp devaluation in the national currency in 1994 which saw a rise in the cost of inputs but the only people benefiting from the increased competitiveness of Senegalese fish were the wholesalers. 80% of landings to the beach in Kayar were bought by the merchants from Dakar (usually represented by just one buyer), the remainder by the local wholesalers. Prices on the beach were fixed but did not change even after the currency devaluation. Fishermen were upset at this situation and, through various local meetings with village chiefs, resolved to take charge of the situation and turn it around to their benefit - and ultimately to the benefit of the resource.

The fishermen went on strike for three days, and when they called off the strike they set up their own distribution chain by renting freezer trucks to ship the fish to the central market in Dakar. After each sale, the profits were shared amongst the fishermen on a pro-rata basis. The determination of the fishermen to improve their lot forced the wholesalers (represented by the principal buyer from Dakar) to enter into negotiations with them. They agreed to raise the fixed price of a box of fish from 700 CFA to 8000 CFA. The agreement did not survive for long however, and the fishermen soon resorted to restricting their catch in order to protect prices by instituting an individual quota, with a set of fines to encourage compliance. The development of the individual quota system has led to increased fisher revenue, decreased fishing time and decreased fishing pressure. The knock-on effect of this on fishing costs has been positive with less intensive use of gear, reduced fuel bills etc.

This new and very successful system draws on local features. The village of

Kayar is difficult to get to, and fishers have little choice but to land there. The village is dominated by one ethnic group with a long fishing history based on line fishing which is far more selective than other gears (especially monofilament nets). Less selective gears have now been banned by the fishers leading to some conflicts with other groups.

Namibia

Namibia is interesting because it is a fairly unique situation. It shows the possibilities that exist for the fisheries sector if this has not been used an employer of last resort. Most fisheries around the world present a difficult socio-political challenge to management because they have been used to generate work. As a result overcapacity is generally a very severe problem, particularly if compared to the economically optimal level of exploitation.

In Namibia where such a policy has not so far been followed, the consequence is that there are very few fishing vessels (only around 300 for the whole country) with very little infrastructure (2 landing points). The consequence is that the fishery not only covers its costs but produces a net return to the national Treasury.

Following independence, the Namibian government was faced with fish resources, particularly hake that had previously been subject to very heavy international exploitation. A first success is that the new government was prepared and able to reduce exploitation rate with the aim of gradually rebuilding stocks. Despite some variation due to environmental fluctuations, the stocks do seem to be recovering.

The most significant element is that following the move from international fishery to lightly-exploited national fishery, significant resource rents have been produced. As a result the debate concerning Namibian fisheries is very different to that concerning most world fisheries, especially in developing countries. In Namibia, the debate turns largely on how to use the benefits of success, i.e. what to do with the rents that have been generated.

In the case study it is argued that too great a proportion of the rents are being invested back into the fishery sector when returns to the economy as a whole appear greater elsewhere. This is an interesting problem that is far from unique to Namibia. Where fisheries ministries and the fisheries sector as a whole is involved in investing resource rents, there is an overwhelming tendency to take a fisheries view of the world. Fisheries ministries tend to focus on developing the offshore and onshore sectors of the industry itself. But there is clearly a need for broad macroeconomic view to be taken, especially if the exploitation of fish resources represents a major source of capital accumulation for the country. It would appear that fish resource rents should not be managed by the fisheries ministry but by the Government as a whole.

Northern Prawn fishery, Australia

The study of the Australian Northern Prawn Fishery illustrates many of the points raised concerning success. The Northern Prawn Fishery is a limited entry, input-controlled trawl fishery. Many economists, including some in Australia, have doubts about the ability of effort-based management systems to produce economic benefits on a sustainable basis because of problems of input substitution. Experience to date with the Northern Prawn Fishery demonstrates that it is possible, through responsive management and industry, to produce benefits for a substantial period, even if a development towards other forms of use right may eventually be needed.

The management of the fishery demonstrates success in a number of areas. The management system has generated net economic returns with substantial amounts of resource rent being earned, capitalised into the value of the licence. Over the 1990s, net returns were usually in the US$ 12-14 million range (on turnover between 80 and 90 US$ million). The capitalised value of licences is estimated at US$ 350 millions. These net returns are over and above management costs which are paid for by the industry.

Managers and all stakeholders have demonstrated a willingness and ability to address sustainability issues. They have been prepared to take the measures necessary, especially effort reduction. As in all input-controlled fisheries, input substitution has been an issue. Management, in collaboration with the industry, has addressed the problem using a range of effort reduction strategies. A major feature has been two major buy-back programmes, costing a total of US$28 million, of which industry paid more than 80%. The schemes reduced the number of vessels by 55% and a broader measure of capacity by 70% by 1993. Capacity was cut a further 15% in 2000 but overfishing continued to be a problem. As a result, industry agreed to reduce capacity by a further 40% in 2002 through a combination of seasonal closures and fishing right reduction.

Prawn trawling is associated with high levels of bycatch and a range of other ecosystem related issues. Responses have included the use of bycatch reduction devices, development of a bycatch action plan and a voluntary but effective ban on the taking of sharks. In addition, extensive closed areas in the Fishery have been established to protect sensitive habitats such as seagrass beds as well as juvenile prawn grounds.

Success is explained by a variety of factors. First, in 1995 existing fishing rights in the Northern Prawn Fishery became statutory fishing rights (SFRs), providing operators with strong, long-term access rights to the Fishery. These rights, combined with industry 'investment' in the fishery through funding for the buybacks, have tended to encourage strong interest by operators in the longevity and sustainability of the Fishery.

Second, since 1992 the Australian Fisheries Management Authority (AFMA) has managed the NPF under a statutory authority framework, at arms length from the political process. AFMA is required to base its management decisions on a set of legislative objectives that include ecologically sustainable development, the precautionary principle and economic efficiency. The NPF management planning process provides a clear 'road map' towards ecological and economic sustainability for the Fishery and includes performance measures against which success in meeting objectives and implementing strategies can be assessed and reported.

A 'partnership approach' has been adopted with a Management Advisory Committee providing advice to AFMA on significant issues and balancing the views of differing industry sectors as well as other stakeholders, including government, research and conservation non-government organisations (NGOs).

Based on the theoretical literature and the empirical evidence from the case studies, the study concluded that there was not one single element that was responsible for success in fisheries management. Factors that lead to success are, broadly speaking:
- cooperation and communication between all stakeholders;
- sufficient institutional capacity (including policy and legislation) to ensure that biological, economic and social objectives are met;
- an ability to deal with the complexity of multiple stakeholders and multiple and often conflicting objectives between stakeholders and related sectors;
- creating appropriate incentives including fiscal systems and use-rights frameworks.

Success in Fisheries Management: A Review of the Literature

Elizabeth Bennett[1]

> Wise men profit more from fools than fools from wise men; for the wise men shun the mistakes of fools, but fools do not imitate the successes of the wise.
> Cato the Elder (234 BC - 149 BC), from Plutarch, *Lives*

Introduction

Lessons in fisheries management all too often start with an examination of what went wrong and then begin to put the problem right. But what if we looked at the successes and attempted to replicate those, rather than always looking at the failures?

The following is a review of how successful fisheries management has been, and can be. The review provided a great many challenges – not least because fisheries managers are either overly modest about their accomplishments and are loathe to document successes where they happen or because good news is simply not worth reporting. However, with concerted effort, as many incidents of success can be found as documentations of failure in the field of fisheries management.

The review is organised as follows. First, a brief methodology of how literature was searched and organised is discussed. Then, success is defined. Success (so too failure) can only be measured against criteria. Success is not an absolute but a relative condition and in order to establish where success has arisen the base-line criteria need to be understood. For the purposes of this review the basic objectives of fisheries management were used as the bench-mark against which success was measured. Section 1 outlines the objectives of fisheries management and examines the nature of success; having defined and identified success, section 2 examines the conditions that lead to success. There are arguably a number of ways that success can be measured, these are discussed in section 3 and section 4 outlines the barriers to success.

Methodology

An initial literature search for evidence of or reference to success produces very little: case-studies demonstrating disaster, failure and human-error abound. Moreover, the overall tone of the FAO Status of the world's Fish Stocks report, an obvious place to find reference to success, is that of resource and capacity problems. What the literature does demonstrate, however, is that stories of success in

---

1   IDDRA, Portsmouth, UK.

fisheries management tend to deal with a number of issues. At first glance, these issues are defined by their quantitative nature (which makes it comparatively easy to justify the claim to success) and the overwhelming emphasis is on biological aspects of the management of fisheries.

First, there is a small literature on individual fisher success which owes much to anthropological research (cf Durrenberger and Palsson, 1983; Palsson and Durrenberger, 1982). Palsson and Durrenberger use quantitative analysis to argue that the supposed links between skipper ability and success were limited. Whilst not directly related to fisheries management per se, they do show that the catch rates of skippers and the factors that may (or may not) influence that are critical to overall success rates of fisheries management programmes.

Second, biological success has come to the fore recently as a number of stocks have been threatened. Coen and Luckenbach (2000) describe the case of Oyster reefs on the East Coast of the USA and Yaragina (1998) describes the case of northeast Arctic Cod. Restocking success draws on cases from both the developing world (eg China: Li, 1999 and India: Reyntjens, 1987) and the developed world (North America: Bachen, 1994) and also deals with restocking of semi-wild species (salmon) and cultured species (shrimp). These cases tend to focus on the economic impacts of such programmes (ie the case of China and shrimp in India).

Finally, the success rates of recreational fishers (see, amongst others Agnello, 1989) receive considerable attention. Whilst the management of recreational fisheries falls outside the realm of commercial fisheries management, these case studies provide some useful measures and criteria for defining success.

The above brief survey of an initial search, however, does little justice to the wealth of information available on successful outcomes of fisheries management initiatives and even less justice to the success that is founded upon qualitative indicators (which are much harder to justify). In order to find these examples, it is necessary to explore the subject from a tangential point of view. Sustainability (of one form or another) is arguably the goal of fisheries management and there are ample examples of sustainable fisheries management from around the world. By adopting a framework that 'measures' success in terms of the ability to achieve objectives, the following analysis will be able to distil evidence of success from the vast literature on sustainability.

If success is simply the difference between the desired and actual outcomes of a given set of objectives, then we need to understand what those objectives are, who set them and how outcomes have been assessed.

Success and the objectives of fisheries management

First, a short analysis of each broad objective is offered. The objectives are defined according to standard categories (biological, social, economic etc), but also with reference to the more comprehensive and explanatory categories identified by Garcia and Grainger (1997:646).

The conservation or bio-ecological objective

Fisheries biology can be conducted without reference to fisheries management, but the opposite does not apply: fisheries managers must take into account the underlying state of the stock. Fisheries are almost exclusively managed for the harvesting of fish, rarely to conserve a species[2] (Hilborn and Walters, 1992: 22). The bio-ecological objective identified by Garcia and Grainger widens the scope somewhat allowing the harvesting and conservation aspects to be combined without prejudice.

It is arguable that fish have an existence value, that is, society places a value on the existence of fish and, if asked to, would be prepared to pay to ensure that fish continue to exist. The same arguments are usually, and successfully, applied to large mammals such as whales and elephants: society is willing to donate money to causes to ensure that whales and elephant populations are conserved because society derives some utility from their continued existence (Bennett and Thorpe, 2003). The argument is rarely applied to fish, yet recent reports that cod stocks in the North Sea are on the point of depletion have seen arguments emerge about why fishing effort should be constrained to ensure that cod stocks are not rendered extinct[3]. The conservation objective, thus, is to ensure that the potential productivity of fish stocks are used to their full advantage without endangering the underlying health of the stock (FAO, 1997). This objective does not imply that no fishing should occur, simply that the level of fishing should be commensurate with the ability of the fish stock to reproduce and maintain a biologically sustainable level. It could be argued that current measures to limit fishing of cod stocks in the North Sea are now being primarily driven by a conservation objective. Ideally, of course, this objective should be the root of any fisheries management plan (without it, the stock may collapse and all other objectives cannot be realised). More often than not, however, the failure of fisheries management results in drastic conservation measures being introduced at the eleventh hour to halt stock decline. See for example Mitchell (1997) on the ban on fishing on the Grand Banks in Canada and Twomley (2000) on the introduction of the first limited licensing programme in the US to cope with the lowest Salmon stock levels in history in 1973.

It could thus be argued that success (or the desired outcome) from a conservation/bio-ecological perspective occurs when the stock (or eco-system) in question is being fished at a sustainable level (be it MSY, F^{01}, MBAL etc). Success can be proven (or argued) on the basis of stock assessment and landings data.

---

2   Except of course where preservation will lead to renewed harvesting as the current situation with cod in the North Sea.

3   Thousands of square miles of cod-spawning grounds are set to be declared off-limits under options being planned by Fisheries Minister Elliot Morley to prevent cod from becoming extinct. Mark Townsend, Sunday December 15, 2002 *The Observer*.

The Economic or techno-economic Objective

A paramount objective for the fishery sector as a whole is to realise its full economic potential, as measured over time by the sum of net economic benefits across all producers and consumers, including rent which could be otherwise extracted" (FAO, 1997:9). In short, the economic objective is to ensure that the fishery is making the most economically rational use of society's resources – that is, that consumer surplus and economic rent are maximised (Cunningham et al, 1985). Additionally, Crutchfield (1965, 1979) argues that the purpose of the fishery is to produce income rather than fish (as per the previous objective) and so costs of catching the fish have to be taken into account. Economic objectives, thus seek to maximise the difference between the costs of fishing and the value of the landed catch thereby ensuring that the resource rent (the value of the catch minus costs of harvesting) is also maximised. Garcia and Grainger (1997) go further, arguing that this objective would be better defined as the techno-economic objective where the goal is to ensure growth and efficiency leading to optimal investment policies and thus fulfilling the long-term duty of care[4].

The most *rational* use of society's resources, however, may not be compatible with a conservation objective; in rare circumstances it may be more rational to mine the resource as fast as possible because the opportunity cost of fishing is too high. It may not be compatible with the social objective (when fishing is a way of life for a community and support for the industry is withdrawn because, economically, it would be better to use resources in another sector). Events in the EU over the past decade have demonstrated that whilst withdrawing subsidies and raising entry barriers to over capitalised fisheries may appear to be the most logical option on paper, in reality the turmoil such measures cause show them to be incompatible with the social objectives perceived by the fishers and to a certain extent the wider public (who place a 'value' on the existence of a fishing industry).

> *Achieving economic success in the fishery requires that fishing effort is maintained at the MEY level. Landings and effort data would demonstrate this adequately, though more sophisticated resource rent capture data would also be needed.*

The Social or socio-cultural Objective

The social objective may be defined as maintaining 'traditional community structure and lifestyles" (Hilborn and Walters, 1992; Anderson, 1986) through employment opportunities and income distribution to areas with little alternative sources of employment. Additionally, it can mean "ensuring equity, participation and empowerment, social mobility and cohesion, designing proper institutional arrangements, taking into account cultural identifies and ethical requirements" (Garcia and Grainger, 1997: 647). The management of the cod fishery in the Lofoten islands in Norway is perhaps a good example. The Lofoten Islands lie off

---

[4] Presumably the long-term duty of care for the economy and the industry held by the Government, though Garcia and Grainger do not elaborate further.

the coast of Norway. They have been dependent upon the Northeast Arctic cod stocks for generations and the tradition of fishing is well established. The fishery is characterised by small operators and governed by ancient laws that ban trawling and stipulate which gears are allowed (most of the fish are caught with hook and line) and the community has ancient ties with export markets across Europe (fishstock is exported to Italy; salted and dried cod to France and Spain). Following the near collapse of the Northeast Cod stocks in the mid-1980s, Norway and the USSR agreed on a set of reference points for the stock in order to help recovery. Norway's objectives were to maintain a sustainable harvest whilst at the same time increasing the profit from the fishery. They also recognised that the fishery was an important employer in rural areas, and thus held the sustainability of rural communities as an equally important objective (Nakken et al, 1996; Urch, 1994). Likewise, following the collapse of the groundfish stocks on the Grand Banks off Canada, management had to modify its conservation objective in order to accommodate the demands for a more social objective as the livelihoods of hundreds of families were affected (Milich, 1999; Allain, 2000; FAO, 2001).

The social objective of fisheries management is perhaps the hardest to define or identify because it can encompass so much. Together with the function of providing employment and nutrition to a community, fisheries also provide links to heritage, a continuation of traditional practices, and a basis for social and community formation. In both developed and developing countries activities associated with fisheries can form the basis for the existence of a community. Fishing communities as different as Gloucester in the United States (Millar and Maanen, 1979) and Chennai in India (Bavinck, 2001) demonstrate that the act of fishing and all the superstition, structure, norms, beliefs and history associated with it are inextricably linked to the lives of those engaged in the industry.

Proving success of social or cultural objectives requires a large amount of data on employment, health, incomes etc. However, the social objective is often largely undefined and requires qualitative measures to show that cultural values are intact, communities are sustainable and a sense of well-being has been achieved.

The Development Objective

Although rarely mentioned in standard fisheries management texts (Cunningham et al, 1985; Charles, 2001) the development objective applies to all governments, irrespective of income, yet is most commonly associated with the less developed world, and this is the sense with which it is used here. The development objective might be defined as the need to use the potential of fisheries to contribute to economic development. That is, fishing becomes a significant rather than secondary driving force in the development process. Moreover, the FAO has asserted that this objective should also include helping LDCs secure their rightful place in world fisheries (FAO, 1984 cited in Royce, 1987:3). The economic objective may however run counter to the development objective. This indeed was the case in many LDCs during the 1970s when substantial sums of public monies and international aid were invested in fishing enterprises and infrastructures (Robinson and Lawson, 1986:101); whilst the economic objective may have been partially achieved, the development objective was largely missed because the large-scale projects failed to deliver real progress to the vast number of small-scale fishers living in poverty and targeting stocks for domestic consumption. The consequences of a mis-match between the economic, social and development objective are most apparent in LDCs where the level of dependency on the resource is often greater due to a lack of government safety nets. Pursuing an economic objective (increased production for export) over a social objective (more equitable access to resources and benefits) can result in the fishery being undermined biologically (through overcapitalisation) whilst the livelihoods of the fishers also suffers as incomes fall. The outcome is that the development objective (the fishery acting as an engine for growth; the fishery acting as a safety net for fishers) is not met. The current situation in Bangladesh amply demonstrates this point: the government is pursuing a biological objective inland (maximisation of fishing production) and an economic objective on the coast (maximisation of rent earned from shrimp exports) yet many of the country's 1 million fishers struggle to make a living from fishing and traditional access to common pool resources is being eroded through increased (illegal) privatisation. (Neiland et al, 2002). Chronic depletion of inland stocks means that the biological objectives are unobtainable; failure to protect access rights or intervene in allocation issues has result in the non-achievement of both the social and development objectives (to improve fishers livelihoods and raise incomes) and on the coast the pursuit of the economic and development objectives has resulted in environmental problems.

Development objectives can be measured on a variety of data: health, mortality, education enrolment, access to basic services, incomes and nutrition. Development success, however, may be a double edged sword: proving success may mean that external aid dries up because it is perceived to be no longer needed and success is thus downplayed. The level at which development has been achieved also provides problems with defining success against this objective.

So, success can be defined as the ability to produce outcomes that meet the stated objectives – or at least come close to fulfilling the objectives. But, objectives and goals are often highly complex; tradeoffs between different view points have to be reached and the interdependence of one factor on another has to be taken into account.

Problems with success
But a number of other crucial questions need to be asked in terms of success.
- Who is establishing the metric for success; in other words, who is defining the objectives and their desired outcomes by which success will be measured? the community may consider the fishery a success (provides employ-

ment) although it may not be meeting management targets (conserving the stocks).
- What might be interpreted as success by one group of stakeholders but be categorised as failure by another – the viewpoint on success and what it means is thus important.
- Success is necessarily a time-bound condition: what is successful under one set of criteria at t_1 may not be regarded as successful under different criteria at t_2.
- If success is the difference between the objective and the ideal outcome then the ideal outcome needs to be stated; if it is not stated there is no way of knowing how well the objective has met its target. In other words, how is success to be identified where weak policy making structures do not identify ideal outcomes?
- Is the achievement of success accepted or it is contested? In other words how is success being measured (qualitatively or quantitatively) and how well does the achievement stand up to questions?
- What is done with the success? Is it capitalised upon, are the lessons disseminated or does it simply go down as another anomaly in the history of fisheries management?

Success is the difference between the expected or ideal outcome and actual outcome, but how might success be characterised?

Arguable sustainability has to be a synonym for success in so far as a fishery that is, for example, biologically or economically sustainable is likely to have achieved one or other of its stated objectives. But what do we mean by sustainability? If a fishery is sustainable then it is able to produce the same results (fish landed, profits made, people employed etc) time and again. Because it is able to repeatedly produce the same results it will be able to pass on these benefits (employment, income, stock etc) to future generations.

Using sustainability as a proxy for success also enables the parameters of success to be widened. In other words, reaching political consensus within a fishery can contribute to sustainability yet political consensus as a marker for success might not ordinarily be considered. Likewise good governance, where decisions are clear and traceable can be considered under a sustainability framework.

But, using sustainability as a proxy for success presents its own problems. Exactly what do we man by sustainability and what are we trying to sustain? Sustainability can be understood as maintaining the *quantity* of natural resources at a consistent level over time or as maintaining the *quality* of natural resources at a constant level over time. In other words, under the first condition future generations will receive no less capital than the current generation but the quality is not guaranteed – capital is assumed to be infinitely substitutable. Under the second condition, however, capital is not assumed to be substitutable and the quality of capital as well as the quantity has to be sustained. Thus, if we understand success to mean that the outcome of a management process is a sustainable fishery, we need to consider whether that success is built upon weak or strong sustainability.

Finally, there is the issue of whether success is tied to equitable or efficient outcomes. Efficient fisheries management systems do not necessarily lead to equitable outcomes; likewise equitable fisheries are rarely efficient (economically or biologically). Catanzano and Cunningham (2001) cite examples from Mauritania, Madagascar and Iceland where the concepts of equitability and efficiency have collided and have impacted upon the reported and supposed success of the fishery. The Icelandic case is the most quoted example of a fishery that is efficient: the introduction of ITQs has seen a reduction in effort and capacity and an increase in rent generated and fish stocks. Yet, the case of Icelandic ITQs is also generally cited as an example of where success achieved through the efficiency route does not lead to success for equity: many sectors of the Icelandic community have grave misgivings over how the quota was distributed and how it has been allowed to be accumulated by a few key operators at the expense of the generalised benefit of the Icelandic nation. It is important, thus to bear in mind that success as seen through one lens may fail to appear as success when seen through another lens.

Measuring Success

Success implies that things have improved, that things are working well, compared to previous situations. In order for the claim of success to be brought two things are needed: first a benchmark on which the improvement can be measured and second a metric to gauge improvement. There are a number of ways of doing this, some less quantitative than others.

Criteria for success

Measuring success requires that some criteria are first established which enable success to be measured 'against' something. Although measurement against management objectives is mentioned above, more sophisticated means of establishing criteria have been developed.

The Marine Stewardship Council, established in 1996 provides a comprehensive set of criteria that not only measure the biological status of the fishery, but, with their expansion into the possible certification of small-scale developing country fisheries, are developing criteria that would measure success (ie the sustainability of the fishery) in social, development and economic terms (Scott, 1997:151).

MSC criteria is largely eco-system based:
- the fishery shall be conducted in a manner that does not alter the age or genetic structure or sex composition to a degree that impairs productive capacity;
- the fishery is conducted in a way that maintains natural functional relationships among species and should not lead to trophic cascades or ecosystem state changes;

but also demands that good governance be maintained
- the management system shall
  - demonstrate [....] a consultative process that is transparent and involves all interested and affected parties so as to consider all relevant information, including local knowledge;
  - be appropriate to the cultural context, scale and intensity of the fishery;
  - provide economic and social incentives that contribute to sustainable fishing and shall not operate with subsidies that contribute to unsustainable fishing.

In May 2003 the Pew Fellows issued a statement on fisheries in response to the increasing concern that fisheries worldwide are being exploited unsustainably. It points out how it thinks sustainable fisheries can be achieved (and thus, presumably what characteristics lead to success in fisheries management and the criteria against which successful fisheries might be adjudged).

Along the same lines as the MSC, the Pew Fellowship establish the following:
- Sustainable fisheries should
  - secure the participation in policy-making and management of all interested parties;
  - establish institutions and forms of governance that provide effective incentives for participants to conserve resources;
  - ensure that allocation systems are establish that provide equitable tenure to domestic fishers;
  - empower consumers to demand fish from sustainable sources.

They identify the following as imperative for fisheries success:
- the elimination of subsidies;
- the provision or identification of alternatives to fishing;
- the maximisation of economic and social benefits from fisheries, particularly those deriving from off-shore fishing rights;
- the setting of targets and regular monitoring of performance;
- the establishment of no-take reserves.

It is interesting to note that the criteria for success in fisheries is often more to do with removing pressure from the fishery (thus allowing it to thrive under less intensive effort) than direct actions to improve the fishery. This stance is reflective of the overwhelming evidence that fishing pressure is excessive and rising and reducing that pressure is the first step needed.

Evaluating a study to restore oyster beds in South Carolina, Coen and Luckenbach (2000) note that criteria for evaluating success need to be establish (and this applies to other stock rejuvenation projects). They suggest that a simple criterion is that economic returns for increased landings together with the social benefits from supporting fishing livelihoods should be more than the cost of renewing the substrate and seeding with oysters. But, they do not offer any means of measuring increased social benefits. They also suggest that target goals for achievement need to be established early in the history of the project – this also addresses some of the information problems described below.

Absence of conflict

The presence of conflict in a fishery is arguably an indication that there is a problem somewhere with the management system. Whilst it is not possible to have a conflict-free fishery, the extent and degree of conflict present is a good indication of how successfully the management regime is able to mediate between the stakeholders and their divergent objectives. Recent work on fisheries management in tropical small scale fisheries suggests that the presence of conflict is a good indicator of how successful the management system is performing and, in the long run, how successful the fishery is likely to be (Bennett et al, 2001).

Measuring success against management targets

Management targets can take a variety of forms. They may be specific limit or threshold points established by fisheries managers; they may be policy statements made by government officials (often, but not always) working in collaboration and on the advise of fisheries managers or they may be sectoral plans (common in centrally planned economies). In extreme cases, the management targets may actually be set or imposed by other agencies – for example in developing countries continued funding may be contingent upon increasing fish consumption among the vulnerable or reducing the use of destructive gears.

Measuring success against such targets assumes, of course, that those targets were correct or achievable in the first place. The case of Bangladesh presents a good example of how inaccurate policies can befuddle attempts to measure success.

A large number of policies have been produced over the last 30 years, covering both national and sectoral agenda; there is considerable overlap and lack of coherence between policies. Poicy-making in Bangladesh has been predominantly top-down originating from central government; the content of policy has favoured the priorities of powerful, elite groups in Bangladeshi society, often at the expense of rural people. What is more, implementation has been extremely variable, and constrained by a weak and bureaucratic institutional setting. The result is that assessment of policy performance and policy evaluation has been limited, with minimal feedback into the policy-formation process. The weakness of the institutional setting and lack of fundamental data have been major constraints. In general, policy performance with respect to development outcomes is considered to be weak, with significant poverty and lack of economic development still a major feature of Bangladesh (Neiland et al, 2002).

Since Independence, Bangladeshi policy has been guided by Five Year Plans (FYP), written by the government[5]. In principal, plans are drawn up to highlight

---

5   With the exception of the Two Year Plan which covered the period 1978-1980.

the goals and objectives that will promote economic and social development and to specify the means of achieving such goals (see Weimer and Vining, 1989 for example). As such, the FYPs give broad outlines to government's intent, reflecting the political priorities and vision of the ruling party.

Targets set for fish production in the 5 Year Plans have consistently failed – not only to meet production figures, but also to spend resources allocated to developing the sector. Yet, further analysis suggests that failure has only occurred in relation to the unfeasible targets.

Several reasons have been given for the failure to meet production targets; 'reduced investment, lack of infrastructural facilities for culture of fish, organisational weakness and environmental effects of irrigation and chemical use in crop production' (3rd FYP); 'lack of technical knowledge, fish seed, proper management, disease control and suitable manpower' (5th FYP); disruption of fish migratory routes, non-compliance with fisheries legislation, short term leasing of open water bodies, lack of good quality data to back up management plans, lack of agency coordination and inadequate programme monitoring and accountability (6th FYP). There are other explanations of why goals for production and spending have not been met: a) they may have been set too high; b) budgets were gradually revised downwards over the period of the FYP to cover shortfalls in other sectors; c) spending did in fact make its target, but not in the fields originally intended by the government (ie, the difference in official figures has fallen under the 'misuse of funds' category)[6]. It could be argued that the FYPs should better recognise the current constraints to development within the sector and realign its targets accordingly. The FYPs have consistently outstripped the capacity of ministries to adapt and achieve targets (Jorgen Hansen (DANIDA), pers. com. April 2002).

Measuring success through scoring systems

It is possible to 'score' success and thus provide quantitative rather than qualitative answers to the dilemma. Arnason (2000) uses a complex formula to score the achievements of ITQ systems in Iceland, Norway and New Zealand. He argues that the formula would allow different systems across country to be compared. He examined both essential and non-essential characteristics of property rights taking four key criteria for successful property rights: exclusivity, security, permanence and transferability. Using 1.00 as an indicator of perfect property rights, he awards New Zealand the highest score at 0.96; Iceland gets 0.86 and Norway 0.44. These scores are based on the ease of transferability (complete in New Zealand, none in Norway); security of tenure (good in all, less so in Norway); permanence (ITQs are issued in perpetuity in New Zealand, indefinitely in Iceland and annually in Norway); and finally exclusivity which is high in all systems. Providing a benchmark for ITQs, Arnason points out that he would

---

6 Keeley (2002) notes that figures quoted in FYP are often of dubious accuracy or 'massaged'.

expect the value of property rights in a CPR to come out at between 0.2 and 0.5. However, in terms of measuring success and the problems of doing this, Arnason then goes on to argue that whilst the system of property rights itself may score well, the quality of the underlying resource is equally important. When this is taken into account, the scores drop considerably: Iceland now has 0.71; New Zealand has 0.80 and Norway has 0.33. Thus, measuring success on the presence of property rights is imperfect because they are ultimately limited by the technical ability to implement them well in the first place. Moreover, success based on this measure fails to account for the social problems with assigning rights to 'public' resources. There is a considerable literature, for example on the problems with ITQs and the manner in which they can lead to concentration of fishing rights in the hands of a few[7].

WHAT IS RESPONSIBLE FOR SUCCESS?

Although all fisheries display different degrees of success and analysis could put forward an endless list of possible factors that contributed to that success, it is possible to characterise success using those factors most frequently mentioned in the literature.

In a study conducted by the OECD in 1997, a range of fisheries management systems from OECD countries[7] were analysed to establish whether a variety of input and output controls and technical measures had met their stated objectives[8]. The following table draws heavily on the findings of the OECD study, with additional factors listed where they appear in the literature. It should be noted that few, if any, of these factors are able to contribute to success on their own: inevitably each factor is operating alongside a number of other factors. The consequence of this of course, is that it is not always possible to pin-point exactly what the cause of success was; the following analysis does, however, indicate the probable role of each factor in the successful management of a fishery. Interestingly, the OECD study covers a wide range of input and output regulations (TAC, individual effort limits, gear and vessel restrictions, limited licensing) and technical measures (size and sex selectivity, time and area closures) but no fiscal measures; only a few are able to demonstrate success (meeting their expected outcomes) – notably those that impact property rights to some degree or other. As would be expected, those regu-

---

7   It should be pointed out that in New Zealand the Maori fought for, and won, the right to 40% of all ITQ in the country thus countering the populist argument (often applied to Iceland) that ITQs are detrimental to small-scale fishing and minority or marginalised groups (Annala, 1996).
8   Only those countries able to present adequate amounts of data, policy frameworks and stated objectives were included in the study; this selection criterion inevitably means that only the more developed OECD countries were included in the analysis.
9   In common with other studies, though, the analysis focuses heavily on the negative impacts of the measures rather than telling the positive and successful outcomes.

Factor	Case-Study	Reasons for success	Source
Use rights	ITQs in Iceland and New Zealand; Canadian Halibut and sablefish fisheries (fishing season lengthened from a few days to nearly 365 days)	Necessary and sufficient to achieve maximum net availability of goods; (partially) resolves prisoner's dilemma; reduces race for fish	Arnason (2000) Annala (1996) OECD (199:113)
Co-management	Japan	Lower transaction costs; stakeholder 'ownership' of the management system; increased compliance; organisational strengthening to absorb changes	OECD (1997)
Compliance and enforcement	Malaysia	Objectives met because legislation to support objectives upheld	Hatcher et al (2000) Kuperan and Sutinen (1995)
Biology and luck	Halibut fishery in NE Pacific	Far easier to manage than multi-species fishery; often with homogenous fleet; Sometimes all the right factors coincide at the right time	OECD (1997); McCaughran (1997)
Policy frameworks	Bangladesh	Provides for attainable objectives and sets ideal outcomes	Neiland et al (2002)
Pro-active industry	NZ implementation of ITQs	Ownership of the process by key stakeholders	Annala (1996)
Stakeholder perceptions	Gulf of Maine Herring Fishery	Local involvement in setting goals that are perceived as fair	Healey (1985)
Multi-sectoral approaches	Philippines Coastal Resource Management Project	Recognises that fisheries have to be managed within the wider context, thus able to deal with potential problems from outside the sector much earlier	Courtney and White (2000)

lations that merely aim to limit catch and/or effort on a global basis rarely achieve their objective.

Stability/equity of Use rights

"Property rights are both necessary and sufficient to achieve the objective of maximising net availability of goods and consequently fundamental to economic progress and well being" (Arnason, 2000:14).

The vast literature on property rights offers many examples of how rights (in which ever form) can have a positive and dramatic impact on fisheries management. Whilst the presence of rights is rarely the sole contributor to success, there is little doubt that the sense of 'ownership' over the resource that derives from rights contributes to a more sustainable pattern of use. And, as Arnason points out, property rights also dramatically reduce the race for fish. Property rights such as ITQs do not guarantee access to a particular collection of fish, but they certainly ensure that only a certain group of users are allowed to target a particular stock, thus reducing the competition for the fish and allowing the fishers to plan their extractive activities more carefully.

There are two types of property that need to be considered here: formal property in the shape of rights that are upheld by law and have 'titles' (in the form of licenses perhaps) and informal property in the shape of rights that are bestowed upon a community through tradition. The former are more common in developed/industrial country fisheries, the latter in developing country fisheries.

The contribution of formal property rights to success is often alluded to in the literature on fisheries management in Iceland and New Zealand. Both fisheries are the most cited and oft quoted examples of successful fisheries management and they are both built upon ITQs. This is not to suggest that ITQs are the most likely form of rights-based management to succeed (there are plenty of detractors to the ITQ argument)[10] rather that this is the form that has been the most appropriate in the circumstances. Arnason (2000) notes that quotas have been successful in both countries in stabilising catch and effort and maximising economic returns and Annala (1996) qualifies this, observing that the participation of the fishing industry in the establishment of such a rights-based system helped it achieve its aims and thus contribute to the greater success of the management goals overall.

The OECD argue that fisheries with a well defined group of users are more likely to have positive outcomes to the introduction of ITQs – mainly because the initial allocation of quota is easier and the fishery has less adjustment to go through – a frequent cause of conflict. (OECD, 1999:115). Yandle (2003) suggests that property rights also form the basis of successful co-management arrangements.

Despite Arnason's assertion above, however, it is arguable that property rights

---

10   See for example Cardenas and Melillance (1999) on ITQs in Chile and Duncan (2002) on ITQs in general.

in and of themselves are not a guarantee of success. Documented experience from Iceland, New Zealand and Namibia (Annala, 1996; Manning, 1998, 1995; Arnason, 2000; Craig, 2000; Oelofsen, 1999) demonstrates that those property rights have to be both equitable in their allocation and seen to be a fair means of controlling effort by the wider group of stakeholders (the nation-state). In other words, whilst fisheries managed under ITQs (for example) may be successful in terms of biological or economic objectives, they may not be so successful in fulfilling their social or development objectives.

Formal property rights require extensive amounts of time and money to set-up and maintain. They also require considerably amounts of data to enable an equitable initial allocation of rights. Developing countries often lack all these requirements which would explain why few such countries have attempted this form of management; Chile (Peña Torres, 1997; Ibarra et al 2000) and Namibia (Manning, 1998, 1995) being the two obvious exceptions to this.

Literature on informal property rights tends to concentrate on examples from the developing world and in particular from common pool resources. The extent of the literature is vast but dominated by such texts as Ostrom (1995); Ostrom et al (1994); Berkes (1986); Pinkerton (1989, 1995). Here, the argument is that informal property rights contribute to more sustainable use of resources, can lead to more equitable distribution of benefits and can thus lead to more successful out comes of fisheries management initiatives. These outcomes may be both economic and biological (as in the case for formal rights above) but also social and developmental.

Co-management arrangements

There is ample evidence from the literature that the active participation of primary stakeholders in the management of fisheries is a key to success (inter alia Fanning, 2000; Pinkerton, 1995, 1989; Pomeroy and Pido, 1995). However, evidence for success is often anecdotal and very hard to back up with biological or economic data.

Co-management implies that management of the resource is shared between the owner (the state) and the user (the fishermen/processors). There is a wide spectrum of variety within the definition of co-management: from 'informational' management (where the state informs the users of decisions made); consultative management (where the state consults with the users before decisions are made) and true co-management (where both state and users take and equal share and responsibility in management) (Jentoft and McCay, 1995).

It is no simple task to reliably credit co-management with specific quantitative success in fisheries management, yet Hanna (1995) notes that in the long term transaction costs are arguably lower when management takes a bottom-up approach. Because the costs of setting up a management system require extensive amounts of information and coordination (the contracting transaction costs) the top-down (state led) management system is cheaper in the design phase. In the long run, however bottom up (user led) management yields better results in terms of compliance and lack of conflict (both of which would normally incur high transacting costs).

Examples of successful co-management can be found worldwide, although it is commonly associated with the developing world and indigenous fisheries in developed countries. Nielsen and Vedsmand (1994) report on the market-based success achieved in Denmark as a result of user, processor and government cooperation. They note that the arrangement successfully ensured quality landings which have helped protect the price. What is more, because there was a high level of identification with the system by the users involved, compliance was high. The management objective was, indeed, to protect and improve the price of the product rather than the resource itself. The reason why this particular fishery has been so successful is that the management boundaries were well defined, the market was well defined, it was concerned with a single species fished by a relatively homogenous fleet. Success in the management of Japanese coastal fisheries is also attributed to co-management. Miki (1997), Taya (1997), Kuronuma (1997), Funakoshi (1995) and Nakanishi (1997) all report on the impact that co-management has had on programmes as diverse as stock enhancement, price maintenance, effort and cost reduction, protecting spawning biomass and growth overfishing. Evidence from Japan also suggests that a social structure that holds conflict avoidance and collective action in high regard lends itself to successful co-management arrangements.

Because of the acute fisheries management problems in Asia (population pressure, poverty) the region has been the focus for co-management initiatives (see Pomeroy and Pido, 1995 for example). Co-management has also proven its worth in contributing to success when combined with other instruments. Because co-management is able to foster collective action amongst users and to strengthen the organisational structure of the fishery it often means that new instruments such as ITQs can be introduced more easily because the system is better able to absorb the changes; this was the case in the Netherlands as reported by Dubbink and van Vliet (1995).

Compliance and Enforcement of laws

With many fisheries management objectives, enforcing the legislation to ensure that effort is reduced, illegal gears are withdrawn and catches are reduced is not only crucial to success but is a key means of measuring how well the objectives are being met. If objectives are only as good as the supporting legislation, then the legislation is only as good as the underlying enforcement mechanisms. Achieving compliance is complex and is often bound by cultural values, economic incentive and perceptions about potential gains and losses (see for example Hatcher et al, 2000 and Neilsen and Mathieson, nd). Kuperan and Sutinen (1995) show how achieving and maintaining compliance was a key to the successful implementation of zoning regulations in the Malaysian fisheries.

Biology and luck

No matter how well defined and executed the management plan, the biology of the underlying resource will be the ultimate deciding factor in the success of the venture. This is often put down to luck (McCaughran, 1997) or to incidental factors

(Sinclair, 1997). Some species respond to fishing pressure better than others: short-lived pelagic stocks are highly susceptible to environmental change (Peruvian anchovy for example) but respond equally fast to revised management plans. Until fairly recently it was thought that demersal species were more resilient to fishing pressure but experience with cod in Canada in particular has dispelled this myth. More encouragingly Yaragina (1998) documents how the northeast Arctic cod stocks were successfully brought back from a state of depletion. Inevitably the coincidence of all the right factors (water temperature, stock health, economic environment, technological capacity) at the right time also has a significant impact on the likely success of a management programme. The contribution of 'luck' to the management equation if amply demonstrated in the case of the Pacific Halibut fishery as documented by McCaughran (1997).

The real difficulty with the biological dimension to success is hat fisheries managers should neither take the blame nor credit for changes due to environmental factors but in practice this tends to be difficult to avoid. Evaluations (formal or informal) of management tend to be based on historical comparisons rather than what the situation would be like now without the management plan or with different measures (that is, the counterfactual position).

Policy frameworks
Management of natural resources has to sit within a national policy framework that guides the management of natural resources. Such a policy framework is made up of the stated objectives of the various state departments (fisheries, trade, environment etc) and the overall macro-economic goals of the government. The strength, flexibility and appropriateness of the framework will have a sizeable impact on the success of the management objectives. Summers (1992) and Meier (1995) cited in Neiland et al (2002) note that the 'super constraint' to national development is poor national economic management; many governments fail to accept sound policy advice (*de facto*) which emphasises sustainable fiscal deficits and realistic exchange rates as a pre-requisite to progress. Successful economic development is a process of cooperation between the state and private enterprise – the problem is how to devise the best possible mixture; a permissive rather than a prohibitive policy environment is essential for the private sector. Good governance is a pre-requisite for effective policy-making and economic development; this means defining the roles of all stakeholders groups and avoiding simple top-down or bottom-up approaches. So, not only is policy important to national development, but to sectoral development too. Whilst much of the work on policy and fisheries has been done in the developing world (this is the point of view that Meier was working to), the same lessons, albeit on a less dramatic scale, apply to the developed world also.

Efforts by fisheries managers to pursue a direction that runs counter to the stated policy will not only lead to frustration, but ultimately will effect the outcome: what success may be evident on the ground will be classed as failure when compared to the policy framework.

Pro-active industry/government

The case of ITQ introduction in New Zealand (Annala, 1996) and rights-based fishing in Namibia both attest to the importance of industry/government cooperation and partnership in the formation of management systems and their execution. The arguments that apply to co-management arguably also apply in this context: that all primary stakeholders need to 'own' the problem and the solution for success to be forthcoming.

Stakeholder (fisher) perceptions

How the principal stakeholders in the fishery perceive the status of the fishery has been shown to have a considerable impact on success. Healey (1985: 173) notes that when fishers believe that fish stocks need supporting, they will agree to measures to limit effort or catch but, if they see a good year class, they will be more inclined to want to fish it out to make up for previous losses. Being able to predict fishing behaviour makes planning and management of a fishery easier; thus being able to predict how fishing behaviour will change is a "powerful analytical tool" for assessing the probable success [and failure] of a particular management tool (Healey 1985: 179). He argues that knowledge of preference structures can make working through compromises much easier. Using the Gulf of Maine herring fishery as an example, he shows that a stock-rebuilding programme will always result in greater overall returns a windfall yield programme, but persuading fishermen that experience a good recruitment year that this is the case is often difficult.

Multi-sectoral Approaches

Courtney and White (2000:42) note that successful fisheries management often has to interact outside the 'fisheries' box; it has to take on board factors that will influence the potential success or failure of fisheries goals such as environmental quality of wetlands and salt marshes, the impact of water-based tourism on fishing activity and structural activities along the coast. Examining the success of Integrated Coastal Zone management in the Philippines, they note that only through the cooperation and interaction of all stakeholders (both within and without the fishing community) have the various management goals achieve positive outcomes. Discussing successful outcomes in broader environmental policy programmes, Vigar and Healey (2002) note that success is dependent upon framing the policy with reference to other adjacent sectors and policy communities. Supporting this remark, Pontecorvo (2003) argues that it is the insularity of those working on 'fisheries', their inability to communicate with related sectors that has partly contributed to the successive failure of management in commercial fisheries.

Need for appropriate Information

Finally information is required to prove that success has been achieved. As pointed out earlier, the ability to measure success (the difference between the desired and actual outcome) usually requires proof – especially in terms of stock size, landings, effort or profits. It also requires that base-line data is available. As Coen and Luckenbach (2000) observe with reference to the restoration of oyster beds in South Carolina, where there is no base-line available on 'pre-project' stock it is impossible to gauge achievements. This sort of data, however, is often missing or contested (cod stocks eg). FAO (1997) note that this is a particular problem in the Western Indian Ocean where a combination of lack of resources coupled with civil war and unrest has resulted in very patchy data collection along the coastlines of the Arabian Sea region. The OECD report on fisheries management provides plenty of examples of successful management systems, but, because it was only able to analyse those fisheries (members of the OECD) with sufficient data, it contains no examples from developing country fisheries

There are two obvious consequences of this: first, that fisheries in LDCs are characterised as failures because there is no concrete proof to the contrary and second, that success is associated with quantitative rather than qualitative achievements thus undervaluing much of the progress achieved in LDCs (social inclusion, co-management and so forth).

However, not all information is appropriate. Where information has been collected as part of a fisheries management plan there is usually sufficient data with sufficient 'purpose' to allow an assessment of the outcomes to be made (indeed a well thought out fisheries management plan relies upon such information being collected). Where information has been collected in an ad hoc fashion it is often less useful; likewise the lack of time series data will not enable any assessment of how sustainable the success is – in other words is the outcome observed a long term trend or simply an aberration in the data set.

The difficulties of trying to assess the performance of fisheries policy in a LDC policy situation have been emphasised recently by Halls & Lewins (1999), as a result of their attempts to research and develop an appropriate methodology. Efforts to develop monitoring and evaluation of national fisheries performance in Bangladesh have focussed on the use of existing data and default indicators to score fisheries outputs against stated national objectives. The DFID supported project "Information Systems for Co-Management of Artisanal Fisheries" briefly reviewed the institutional and methodological limitations of existing data collection strategies such as the Fisheries Resources Survey System (FRSS)[11]. The FRSS effectively represents all catch statistics available for planning, development and management but there has been consensus that the system is outdated, inaccurate and insufficiently resourced. In addition to the technical and statistical limits of the FRSS there are also fundamental institutional and political issues that relate to its function.

This a particular problem where fisheries not perceived to have a value, or their value is far below other economic sectors (inland fisheries in Latin America for

---

11 The FRSS was developed with the assistance of the FAO/UNDP (BGD/79015) project between 1980 and 1984 to provide reliable fisheries statistics for riverine, beel, baor, pond, floodplain, Sunderbans and marine systems.

example) or where fisheries management plans are not evaluated or monitored (Agostinho and Gomes, nd: 8).

Finally, information distribution is also subject to the ability to communicate. As Pontecorvo notes (2003), academic disciplines engaged in fisheries research are not necessarily good at communicating information across disciplines (oceanography, biology, economics, sociology) which frustrates the distribution of information.

Conclusions

Despite the doom-laden pronouncements of some academics and many environmental organisations, there are plenty of examples of fisheries that can demonstrate success over a wide range of criteria. There are biologically successful fisheries where species have been rescued from near stock-depletion; there are fisheries that have been successfully turned around from over-capitalised ventures to economically viable sectors able to demonstrate full cost recovery. Fisheries that have been able to support communities over long periods of time are in evidence in many parts of the world as are fisheries that have succeeded in ensuring that the benefits have been equitably distributed. However, because fisheries management science has essentially been a 'problem solving' discipline, it invariably looks for the problem, oblivious to the successes in a particular system.

Part of the problem, of course, is the difficulty of defining success and, once defined, measuring it. Where success has been recognised it has often been associated with biological and technical factors (because these are easy to measure) with little credit given to success that is derived from and characterised by institutional, social or economic factors. Because the ability to 'measure' success is so critical to its acceptance and credibility, there are far more examples of successful fisheries management from the developed world (which benefits from advanced data collection systems, extensive legislation and controls) than there are from the developing world. This is not to suggest that fisheries in the developing world are unable to demonstrate success, rather it suggests that the means of measuring success are still too quantitative to accommodate the very different context that dictates how fisheries are managed in the developing world. Reference has already been made to attempts by the Marine Stewardship Council to adapt current criteria for defining sustainability to developing country fisheries; those same criteria need to be adopted by fisheries analysts to enable lessons of successful and sustainable management in the developing world to be communicated and disseminated to the developed country fisheries.

The problem with success, however, is holding on to it. Success is not a constant but a dynamic state – largely based on the status of the underlying stock and it is often not easy to continue to replicate success and prevent it becoming simply an anomalous blip in the data rather than an on-going trend. Monitoring and an internal feedback mechanism are clearly needed to ensure that success, once attained is maintained. Yet, monitoring systems are costly to implement and are often the preserve of developed countries.

Finally, it should be remembered that success can be the cause of failure: it can act as a powerful magnet for the fishery. As news of increased catches, increased profits spread so there is a danger that more effort will enter the fishery. To prevent success subsequently turning into failure, it has to be carefully managed and the consequences of improved conditions have to be well mapped out and catered for in advance. In short, with appropriate and well-devised fisheries management plans in place, success should be the inevitable outcome for the fishery and the management plan will provide adequate support to ensure that success is sustainable.

References

Agnello, R.J. 1989: The economic value of fishing success – an application of socioeconomic survey data. *Fishery Bulletin.* 87 (1): 223-232.

Agostinho, A.A. and Gomes, L.C.: Biodiversity and Fisheries Management in the Parana River Basin: Successes and Failures. Universidade Estadual de Maringa, Brazil. Unpublished manuscript.

Allain, M. 2000: Co-management the way forward. *Samudra*, 25 April 2000.

Anderson, L.G. 1986: *The economics of fisheries management, reviewed and enlarged edition.* Johns Hopkins University Press: Baltimore, Maryland.

Annala, J.H. 1996: New Zealand's ITQ system: have the first eight years been a success or a failure? *Reviews in Fish Biology and Fisheries* 1996: 6: 43-62.

Arnason, R. 2000: Property rights as a means of economic organisation. *Use of property rights in fisheries management,* T404/1 FAO: Rome, pp. 14-25.

Bachen, B. 1994: The Impacts of success: a case history of Hidden Falls hatchery. In: *Northeast Pacific Pink and Chum Salmon Workshop.* Fairbanks, Alaska: Alaska University, Sea Grant Program. Pg 47-56.

Bavinck, M. 2001: *Marine Resource Management: conflict and regulation in the fisheries of the Coromandel Coast.* Sage: New Delhi.

Berkes F. 1986: Local-level Management and the Commons Problem: a Comparative Study of Turkish Coastal Fisheries. *Marine Policy,* 10 (3), 215-229.

Cardenas J.C. and Melillanca P.I. 1999: *Individual Transferable Quotas: The other side.* Samudra, Issue 22.

Catanzano, J. and Cunningham, S. 2001: Equity and management instruments in fisheries. Paper presented at a conference in Aix-en-Provence, September 2001.

Charles, A.T. 2001: *Sustainable Fishery Systems.* Blackwell Science: Oxford.

Coen, L.D. and Luckenbach, M.W. 2000: Developing success criteria and goals for evaluating oyster reef restoration: ecological function or resource exploitation? *Ecological Engineering,* 15 (3-4): 323-343.

Courtney and White. 2000. Integrated coastal management in the Philippines: testing new paradigms. *Coastal Management* 28: 39-53.

Craig, T. 2000: Introducing property rights into fisheries management government cannot cope with implementation alone. *Use of property rights in fisheries management,* T404/1 FAO: Rome pp. 17-22.

Crutchfield, J.A. 1979: Economic and Social implications of the main policy alternatives for controlling fishing effort. *J. Fish. Res. Bd. Can.* 36:742-752.

Crutchfield, J.A. (ed.) 1965: *The Fisheries Problems in Resource Management.* University Washington Press: Seattle.

Cunningham, S., M.R. Dunn and D. Whitmarsh. 1985: *Fisheries Economic and introduction.* Mansell Publishing Ltd: London.

Dubbink, W. and M. van Vliet. 1997: *From ITQ to co-management? Comparing the usefulness of markets and co-management illustrated by the Dutch flatfish sector.* OECD. Towards Sustainable Fisheries, Issue Papers.

Duncan, L. 2002: Global economy, global fisheries? *Samudra,* Issue 33.

Durrenberger, E.P. and Palsson, G. 1983: Riddles of Herring and Rhetorics of Success. *Journal of Anthropological Research,* 39 (3) pp. 323-335.

Fanning, L.M. 2000: The co-management paradigm: examining criteria for meaningful public involvement in sustainable marine resource management. In: Mann Borgese, E.; A. Chircop; M. McConnell and J.R. Morgan (eds): *Ocean Yearbook No. 14.* The University of Chicago Press: Chicago and London.

FAO. 2001: Country Profile: Canada. http://www.fao.org/fi/figis/fishery/demos/countryprofiles/FIMS_CAN_E1.htm (last accessed 26 June 2003).

FAO. 1997: *Technical guidelines for responsible fisheries management 4.* FAO:Rome.

FAO. 1997: Review of the State of the World Fishery Resources: Marine Fisheries. Fisheries Circular No. 920 FIRM/C920. FAO: Rome.

Funakoshi, S. 1995: Case Study of Reproduction-Oriented Stock Management in Governor-Licensed Fisheries: Sand Lance Stock Management in Self-Management and Co-management of Coastal Fisheries in Japan, OECD. OECD Towards Sustainable Fisheries: Issue Papers.

Garcia, S.M., and R. Grainger. 1997: Fisheries Management and Sustainability: A New Perspective of an Old Problem? In: *Developing and Sustaining World Fisheries Resources: The State of Science and Management: 2nd World Fisheries Congress,* ed. D.A. Hancock et al., 631-654. Collingwood, Victoria, Australia: CSIRO Publishing.

Hannesson, R. 1993: *Bioeconomic Analysis of Fisheries.* Fishing News Books: Oxford.

Hatcher, A., S. Jaffry, O. Thebaud and E. Bennett. 2000: Normative and social influences affecting compliance with fishery regulations. *Land Economics.* 76 (3): 448-461.

Hilborn, R. and C.J. Walters. 1992: *Quantitative fisheries stock assessment: choice, dynamics and uncertainty.* Chapman and Hall: London/New York.

Ibarra, A.A; C. Reid and A. Thorpe. 2000: Neo-Liberalism and Its Impact on Overfishing and Overcapitalization in the Marine Fisheries of Chile, Mexico, and Peru. *Food Policy.* 25(5): 599-622.

Kuperan, K. and J.G. Sutinen, 1995: Compliance with Zoning Regulations in Malaysian Fisheries. In: *International Cooperation for Fisheries and Aquaculture Development: proceedings of the VIIth Conference of the International Institute of Fisheries Economics and Trade,* Vol. I, ed. D.S. Liao, Keelung, Taiwan: National Taiwan Ocean University.

Kuronuma, Y. 1997: Case Study of Fishing Ground Management in Unrestricted Fisheries: Self-Imposed Management of Vertical Long Line Fishery in Alfonsino Fishing Ground off Katsura, Chiba Prefecture. *Towards Sustainable Fisheries: Issue Papers.* OECD.

Li, J. 1999: An appraisal of factors constraining the success of fish stock enhancement programmes. *Fisheries Management and Ecology.* 6, 161-169

McCaughran, D A. 1997: Seventy-five years of halibut management success. In *Developing and Sustaining World Fisheries Resources: The State of Science and Management: 2nd World Fisheries Congress,* ed. D.A. Hancock et al., 631-654. Collingwood, Victoria, Australia: CSIRO Publishing.

Manning, P. 1995: Managing Namibia's Fisheries as a Vehicle for Development: A Common Pool Resource or Candidate for Tradable Pool Rights? Paper presented at "Reinventing the Commons," the fifth annual conference of the International Association for the Study of Common Property, May 24-28, 1995, Bodø, Norway.

Manning, P. R. 1998: *Managing Namibia's marine fisheries: Optimal resource use and national development objectives.* PhD thesis, London Schoool of Economics and Political Science, London.

Miki, K. 1997: Case Study of Farming Stock Management in Fishing Right on Common Fishing Right Fishery – Scallop Fishery Management in Sarufutsu Area in Hokkaido. In: *Towards Sustainable Fisheries: Issue Papers,* OECD.

Milich L. 1999: Resource Mismanagement Versus Sustainable Livelihoods: The Collapse of the Newfoundland Cod Fishery *Society and Natural Resources* 12 (7): 625-642

Millar M.L. and J.V. Maanen. 1979: Boats don't fish, people do: some ethnographic notes on the federal management of fisheries in Gloucester. *Human Organisation.* 38: 377-385.

Mitchell, C.L. 1997: Fisheries management in the Grand Banks, 1980-1992 and the straddling stock issue, *Marine Policy,* 21(1):97-109

Nakanishi, T. 1997: Recruited Stock Management and Farming Stock Management in Governor-Licensed Fisheries and Minster-Licensed Fisheries Case Study – Fishery Management of Olive Flounder (Paralichtys Olivaceus). *Towards Sustainable Fisheries: Issue Papers*. OECD.

Nakken, O; P. Sandberg and S.I. Steinshamn. 1996.: Reference points for optimal fish stock management: a lesson to be learned from the Northeast Arctic cod stock. *Marine Policy*. 29 (6): 447-462.

Neiland, A; E. Bennett and R. Lewins. 2002: *Fisheries Sector Review and Future Development. Theme Study: policy frameworks*. Report produced for DFID: Dhaka.

Nielsen, J.R. and G. Mathiesen: Important Factors Influencing Rule Compliance in Fisheries - Lessons from Danish Fisheries. Paper 51 Institute for Fisheries Management and Coastal Community Development (IFM): Denmark.

Oelofsen, B.W. 1999: Fisheries management: the Namibian approach, *ICES Journal of Marine Science*, 56 (6) 999-1004.

Ostrom, E. (ed.) 1995: *Governing the Commons: the evolution of institutions for collective action*. Cambridge University Press: Cambridge.

Ostrom, E., Gardner, R. Walker, J. (eds.) 1994: *Rules, games and common-pool resources*. Michigan: University of Michigan Press.

Pena-Torres, J. 1997: The Political Economy of Fishing Regulation: The Case of Chile. *Marine Resource Economics*. 12(4): 253-280.

Pinkerton. E. (ed.). 1995: *Fisheries that work: sustainability through community-based management: a report to the David Suzuki Foundation*, Vancouver.

Pinkerton, E. (ed) 1989: *Cooperative management of local fisheries. New directions for improved management and community development*. University of British Colombia Press: Vancouver.

Pomeroy, R.S. and M.D. Pido. 1995: Initiatives towards Fisheries Co-management in the Philippines, the Case of San Miguel Bay. *Marine Policy* 19, 3: 213-226.

Pontecorvo, G. 2003: Insularity of scientific disciplines and uncertainty about supply: the two keys to the failure of fisheries management. *Marine Policy*. 27: 69-73.

Reyntjens, D.J.R. 1987: Confined-pond shrimp farming near Chilka Lake: a success story from Orissa on small-scale shrimp pond culture. *Bay of Bengal News*. 25: 8-10.

Robinson, M.A. and R. Lawson. 1986: Some reflections on aid to fisheries in West Africa. *Marine Policy*. April: 101-110.

Royce, W.F. 1987: *Fishery Development*. Academic Press: Florida, USA.

Scott, P. 1998: Principles and Criteria for Sustianable Fishing. *Marine Environmental Review of 1997 and Future Trends*, 5 (21): 151-154.

Sinclair, M. 1997: Why do some fisheries survive while others collapse? In: *Developing and Sustaining World Fisheries Resources: The State of Science and Management: 2nd World Fisheries Congress*, ed. D A Hancock et al., 163-166. Collingwood, Victoria, Australia: CSIRO Publishing.

Taya, K. 1997: Case Study of Fish Price Maintenance Oriented Management in Governor-Licensed Fisheries: Self-Management of Mantis Shrimp Fishery in Yokohama. *Towards Sustainable Fisheries: Issue Papers*. OECD.

Twomley, B. 2000: *Commercial fishing license limitation in the State of Alaska: a controversial system of grandfather rights*. Paper presented at IIFET 2000, July 10-14, Corvallis, Oregan, USA.

Urch, M. 1994: 50,000 ton island cod fishery,. *Fishing News* June 19(4196): 10-11.

Vigar, G. and P. Healey. 2002: Developing Environmentally respectful policy programmes: five key principles. *Journal of environmental planning and Management*. 45 (4): 517-532.

Weimer, D.L. and A.R. Vining. 1989: *Policy Analysis, concepts and practice*. Prentice-Hall International, Inc: London.

Yandle, T. 2003: The Challenge of Building successful stakeholder organizations: New Zealand's experience in developing a fisheries co-management regime. *Marine Policy*. 27 (2): 179-192.

Yaragina, N.A. 1998: Assessment and management of northeast arctic cod: failure or success? In: Eide, A. and Vassdal, T. (eds): *IIFET '98 Tromsø, proceedings of the ninth biennial conference of the International Institute of Fisheries Economics and Trade*. Volume 2. University of Tromsø, Norwegian College of Fishery Science: IIFET; 1998: 663-668

Management of the Pacific Halibut Fishery

John M. Gates[1]

Introduction

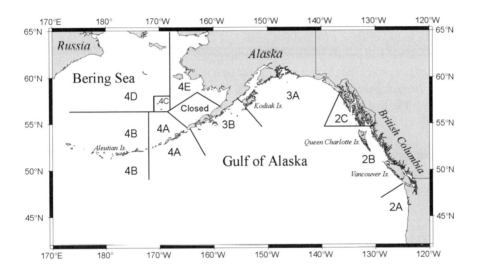

The Pacific halibut (*Hippoglossus stenolepis*) lives in the marine waters of the North Pacific along the continental shelf from Southern California to Alaska and along the coasts of Japan and Russia[2]. The species is considered one of the largest flatfish (second only to the nearly extinct Atlantic Halibut). The maximum age of halibut is about 40 years at which age they may weight more than 200kg and exceed 2m in length. Most halibut captured, however, are only 8-15 years of age and their weight is in the (4.5, 45) kg range. Females are typically larger than males and almost all fish of 45^+ kg are females, with few males exceeding 35 kg. The Pacific halibut is demersal, i.e., it is a bottom-dwelling fish, found over such varied bottom types as mud, sand, or gravel banks at depths up to 1100 meters. However, most are caught at 2 to 25 percent of this maximum reported depth. Halibut migrate to

---

1   University of Rhode Island, USA.
2   Halibut acquired its common name from "haly-butte" in Middle English, meaning flatfish to be eaten on holy days.

deeper waters (200-450m) during winter and, after spawning, they return to shallower coastal areas where they feed for the summer. In addition to their annual migrations inshore-offshore, halibut exhibit age-dependent migration tendencies. As they age, they tend to migrate to deeper waters and to extensive geographic migrations which tend to offset a natural drift of eggs and larvae in a North-easterly direction.

Gear and Vessels

According to DesHarnais (undated)

> "The halibut fishery was and is a hook and line fishery. Early attempts at trawling for halibut had proved unsuccessful. Fishing gear was set and retrieved using either small two-man dories, which were carried by a mothership, or else by "longliners" that set the gear behind them. By the 1930's, dory fishing was no longer practised and longlining had become the standard method. Halibut fishing gear was standardised early in the history of fishery".

The gear type is relevant to subsequent discussions but further discussion of gears seems unnecessary at this time. However, there have been significant productivity changes, to which we will return as needed.

Markets

Marketing services include moving commodities between supply and demand points in space. Marketing involves, also, transformation of raw product to finished product. Such transformations for fish may involve heading and evisceration, filleting freezing, breading, packaging, etc. Marketing may also involve moving commodities to a demand point later in time than that of its landing. Thus, the marketing of products has been characterized as "markets in space, form and time". The transformations in space, form and time are strongly influenced by technological changes in transportation (e.g. railroads) and food science, for example, Clarence Birdseye's development of frozen fish sticks. Other factors in changing markets involve regional supply (e.g. depletion of Atlantic halibut stocks) and demand shifts (e.g., shifts in consumer tastes and preferences toward fish-based products). The decisions about transformations in space, form and time are driven by market calculations of marginal revenues and costs; change of product form occurs if the value added exceeds the cost of transformation. Crutchfield and Zellner (2003, p. 45), estimate the costs of freezing at about 4.4 percent of product price and storage costs, including interest costs at about 0.7 percent of product price. Product flows to a new geographic market only if the revenue gain exceeds the transport costs and if the net margin is at least as great as in other markets[3]. The reverse of this is that a potential new producing region receives an effective demand only if the conversion, storage and transport costs differ from the wholesale price by at least the cost of harvest. Storage and transfer of products to a future market is expected only if the discounted future price exceeds the current price by at least the cost of product conversions and storage costs. For raw product with a pronounced seasonal cycle, the approaching new season places strong incentives to sell off durable (e.g. frozen) inventory before the influx of new seasonal supplies.

In a regulated fishery, these calculations can be distorted in several ways. Product conversions can increase product durability (e.g. frozen vs. fresh fish). The existence of a durable product form connects inter-temporal markets that in earlier days could be connected only via decisions to postpone harvests of fresh product. The latter is a conservation decision based on rational economic calculation; the former may not be. In the absence of exclusive rights to harvest, this inter-temporal connection of fresh markets is null and void; the race to fish is limited only by the marginal costs of harvest and contemporaneous price. Development of frozen product forms can worsen the problem of economic waste if fishers lack secure use rights. Markets in space, form and time become connected in unnatural ways. Because the frozen product is durable, it can be stored for lengthy periods. Given the non-exclusivity of fishers' use rights, this seems like a gain. But this private perception of gain can involve a social loss in that a superior product form may be converted to an inferior one, consumers suffer a loss in consumers' surplus and unnecessary costs of storage are incurred. That quality problems existed in halibut markets is illustrated by the following quote from Crutchfield and Zellner (*op. cit.*, p. 46):

> *"A significant number of individuals in all phases of the industry have expressed concern over the quality problem. Several out-of-state dealers interviewed stated specifically that the competitive position of halibut relative to... (substitute species)... has been weakened in recent years because of uneven quality. Frozen fillets, in particular, can be handled, processed and marketed very rapidly because groundfish are landed regularly throughout the year, and since the packaged items are normally branded, control of quality becomes mandatory for continued buyer acceptance."*

Notice the evident importance of uniform, predictable seasonal flows of raw product in this market. We will see below that these concepts of markets in space, form and time have important implications for the unintended (mis)management of fish such as the Pacific halibut.

Product form preferences and associated price premiums are not immutable, however. A price premium for fresh fish is influenced by relative scarcity. Frozen product forms are more durable and offer flexibility in the marketing chain which

---

3   Which leads to an equilibrium inference: prices for an homogeneous product in spatially isolated markets tend not to differ by more than the cost of transport.

a fresh product does not. A vendor of fresh product is in a risky position if he holds fresh product for which he does not have either a pre-existing contract (written or verbal) to buy. In general, markets are risk averse and if fresh product supplies increase suddenly, old price premiums may narrow or even vanish. This point is nicely illustrated by the following quote from Crutchfield and Zellner (*op. cit*, p. 44-45):

> "During the depression, Washington's share shot up, largely because prices were so severely depressed that transport charges from the northerly ports could not be met."

An implication would seem to be that, *ceteris paribus*, a substantial increase in supplies which depressed fresh halibut prices could also make it difficult for Alaskan producers to compete with British Columbia fishers in the fresh product market.

HISTORY OF THE FISHERY [4]: Pre-1923[5]

Atlantic halibut, the market predecessor of Pacific halibut, was initially regarded by New England fishers, as a pest which drove their target species (cod) away from their lines. For later generations of fishers, (by 1820 -1825), the merits of halibut as a food were appreciated and, as their price rose, New England fishers went as far as the Grand Banks of Newfoundland to harvest halibut for eastern markets. Quality was in part, a function of distance and by the mid-nineteenth century, fish from areas nearer markets commanded higher prices. While market proximity ensured a quality and price premium, it also ensured the early depletion of such stocks. With depletion of Eastern stocks, the situation was ripe for discovery of a substitute species and the means to bring it to market in a quality state. The opening of the Northern Pacific Railway provided the means, and barely exploited stocks of Pacific halibut became the substitute species. Thus, the Pacific halibut fishery is much younger (last decade of Nineteenth century), than its Eastern counterpart. From inception, the Pacific halibut fishery has been a deep sea, international (US and Canada) fishery, principally in extraterritorial waters.

The Pacific halibut fishery followed a trend that has become all too familiar in fisheries. Depletion of substitutes near markets led to higher prices. Improved transportation technology brought new stocks within the market zone of one or more major markets. Fishers responded to the increased ex-vessel demand by building ever more vessels, at ever increasing fishing power and range of operation. This in turn expanded the zone of exploitation as localized depletion made travel attractive to ever more distant waters. As long as new stocks became prof-

---

4   Source: http://www.pac.dfo-mpo.gc.ca/ops/fm/Groundfish/Halibut/history.htm
5   Source: http://www.iphc.washington.edu/halcom/history/pretreaty.htm

itable to exploit, expansion could, and did, continue without much worry about resource depletion. By 1910, the fishery extended over some 3000 kilometres from Northern California to the Bering Sea. Eventually, in 1914, controls on the fishery were proposed, not to conserve the resource but to prevent overproduction and depression of market prices. Even so, it was realized that the stocks of halibut on the older banks were being reduced. Barely a decade had passed since the infamous statement by T.H. Huxley (who served on three British fishing commissions), argued that herring and cod could never be fished out – nature, being inexhaustible in the Victorian view. The intellectual basis for understanding resource depletion simply had not yet evolved.

Despite the evidence of localized depletion, the habit of opening new, more distant grounds delayed any serious recognition of the larger picture. A closed winter season was proposed circa World War I. Still, the rationale for the closure was not conservation per se, but to reduce fishing during a dangerous season, when many fish were spawning and in poor condition for the market, and to reduce supplies while frozen inventories were liquidated. This was to prove a fateful step as it set the stage for a classic tale of economic folly; a marvellous example of the myopia of a command and control management system which, due to an imbalance of use rights, is uninformed by any larger vision of the system being manipulated[6]. The first two decades of the 20th century were watershed periods in the history of the North American Conservation movement; but it was only a beginning. Reports by the biologist W.F. Thompson in 1916 and 1917 were the beginning of a formal recognition of the biological problem. Shortly thereafter, editions of the Report of the Commissioner of Fisheries for the Province of British Columbia showed a sharp decline in halibut on older grounds and the need for control of the fishery. Conservation began to be stressed more and more as the object of proposed regulations rather than a peripheral topic.

Management Strategies 1923-1980

The 1923 Convention and the IPHC

The Canadian fishery is about 10 percent as large as the US fishery, while within the US, the halibut fishery is overwhelmingly dominated by Alaskan landings. Thus, we will focus on the Alaskan and Canadian components of the fishery; the former because of its size; the latter because that is where the transition to rights-based fishing first occurred in the halibut fishery. Conflicts between the two countries led to the creation, in 1923, of a US-Canadian Convention and creation of an International Fisheries Commission (later to become the International Pacific

---

[6] This is too hard on our predecessors; their terms of reference did not include economics and the necessary concepts had not yet been developed.

Halibut Commission (IPHC)[7]. The 1923 Convention has been amended several times. The Canadian Coastal Fisheries Protection Act extended Canada's fishery jurisdiction to 200 miles from shore, beginning in 1977. In 1979, the Protocol to the Convention of 1953, signed by the two countries, brought an end to U.S. fishing in Canadian waters. The Protocol also enabled the individual governments to make regulations that did not interfere with IPHC regulations, pertaining to their own fleets. Canada immediately limited entry into its halibut fishery in 1979. In subsequent years, we have learned of the problems posed by limited entry; especially the effects of input stuffing and the race to fish, but in 1979, limited entry seemed a great step forward. Some comments at the time, suggested vessels were sufficiently inflexible that input stuffing would not be a problem. This may reflect confusion between technical flexibility and cost inflation. While large sums of money may be spent on input stuffing with only modest changes in fishing power, it is the cost inflation which reflects economic waste. There can be "normal" technological progress, which tends to lower unit cost. There also can be "induced" (by regulations) technological progress, and associated input stuffing, which tend to raise unit costs of harvest without much change in catch. Burnes (1998, p. 1) indicates that the introduction of circle hooks, beginning in 1983, were later shown to have increased fishing power by 39-61 percent. There is no indication whether this was normal or induced technological progress[8]. All that can be expected as a limit on the process is that the ratio of unit input cost to incremental output be less than ex-vessel price. One possibility is the soak time concept of trap fisheries. If the number or length or design of lines is restricted, it becomes rational to recycle gear more often - to reduce the soak time (Gates, 2000). In a trap fishery, reducing soak time can be expensive in terms of labour, bait, fuel etc. On the other hand, use of excessive amounts of gear can be a means of "pre-empting territory " and can result in excessively long soak times; so long that the quality of fish recovered is poor. Desharnais' discussion indicates a long line soak time of 12 hours and substantial technological progress in the first three decades of the 20th century, but seems to imply a static situation thereafter. I found little discussion of changes in soak time, or what such changes would imply for operating costs. This period was a period of social experimentation by biological managers of the IPHC. DesHarnais (*op. cit.*, p. 3) provides the following succinct statement:

> "*The regulatory history of the halibut industry provides an opportunity to test the consequences of the Commission's policy. The period 1928 to 1960 is of particular interest as it provides both reliable data and a continuous period of regulation at the end of which, according to the statistical analyses of the Commission, the biological goal of the greatest maximum sustainable yield was reached.*"

---

7   1923 Convention for the Preservation of the Halibut Fishery; subsequently. replaced by the 1959 Convention for the Preservation of the Halibut Fishery of the Northern Pacific Ocean and Bering Sea. http://www.oceanlaw.net/texts/iphc23.htm
8   Discoveries can be motivated by perverse incentives, but once discovered, knowledge is indivisible and may not be economically reversible, even if the perversities disappear.

Yet, all was not well. In the closing years of this period, economists entered the debate. Students of fisheries economics are familiar with H. Scott Gordon's classic 1954 paper and the analysis of the Pacific Halibut fishery by Crutchfield and Zellner (*op. cit*). They pointed out that the Commissions policies had failed economically and could only be considered half done. The reaction of the IPHC staff was defensive and dismissive and H. Scott Gordon was publicly chastised by Commission members (DesHarnais, *op. cit*). From the Commissioners' perspective (biological conservation), their task had been well and truly discharged.. Again, quoting DesHarnais:

> *"One-time Commission member F.H. Bell's seminal history of the Pacific halibut industry was also dismissive of the economists. Bell argued they prejudged the Commission's regulatory success by relying on "hearsay" to confirm their theories about "the common property character of the fisheries generally and of the Pacific halibut fishery in particular.... Bell makes two claims that contradict the fishery economists' conclusions of over-capacity. First, he argues 'there was little evidence of fleet increases tying up resources to any consequential degree.' Second, 'by 1965 all productivity measures per crew member had about doubled over what they were immediately prior to regulation.'*

His response is profoundly ignorant of the economic concept of productivity and the potentials for input stuffing[9].

Gordon is quoted in transcripts of the expert panel with the following smooth response:

> *"GORDON withdrew his strictures on the Commission. They have done a good job. The complaint must be against the economists and others who did not succeed in having the Commission's terms of reference made wider."*

The Pacific Halibut fishery is sometimes regarded as a model fishery in terms of biological conservation but a failure in terms of economic objectives such as cost effective harvesting. We will have more to say on cost effectiveness, but a recent paper by Laukkinen (2000, under review), enables us to address the conservation issue, at least from an economic perspective. The paper also gives insight into the potential gains from rights-based fishing under the assumption that said rights are sufficient to achieve the potential. The Laukkinen paper uses several modern estimation methods, including Generalized Method of Moments and Empirical Moments Method (as well as more traditional methods such as ordinary and non-linear least squares), to estimate the Euler equations of an omnipotent omniscient

---

9   (1) If you raised your prices would it increase your profits? (2) If you lowered your prices would it increase your profits? (3) Do you set your prices so as to maximize profits? In a classic survey, producers overwhelmingly responded "no" to all three questions.

central fishery manager. It is not difficult to find reasons why the modus operandi of managers in any fishery may have differed from such an abstraction. Nevertheless, this austere representation of the fishery is useful precisely because of its austerity. One could use as a reference point, the stock level expected in the absence of management. Successful biological conservation can always be claimed if one has a sufficiently pessimistic expectation for laissez-faire, open access. The oft-cited candidates for regulation are Maximum Sustainable Yield (MSY) and Maximum Economic Yield (MEY). A zero discount rate would make MSY the economic optimum. The appropriate discount rate for an economic optimum is usually thought to be in the range of 5-10 percent. A higher discount rate implies an even lower state of conservation. In the Pacific Halibut Fishery, during the period 1935-1977 (structural changes in data systems make post 1977 comparisons difficult), Laukkinen found an implicit discount rate of 150-163 percent and estimates that profits could have been doubled had the exploitation rate been decreased (escapement rate increased) to an economic optimum. Empirical work always involves a degree of imprecision, and she did bring to bear more powerful methodologies than her predecessors. Nevertheless, one must conclude that this fishery was not well managed in terms of optimal exploitation rate or MSY. I regard this conclusion not as a criticism on the quality of the work done by the IPHC staff, but rather the practical impossibility of achieving biological and economic objectives without the help of rights based fishing. At least as important as the exploitation rate is the cost-effectiveness in achieving this rate. On these grounds it will be seen that management also left much to be desired. The introduction of rights-based fishing has gone far toward the cost-effectiveness criterion and recent increases in yields are suggestive of improved conservation as well. Other important objectives have also been advanced by rights-based fishing.

Management Strategies 1980-1995

Canada

Although the Canadian fishery is much smaller than the US fishery, it would be narrow -minded to ignore events across the common border which is permeable to the resource and to market flows. Also, Canadian decision-making with respect to new ways of managing preceded similar actions in the US and the success of earlier Canadian measures undoubtedly played a role in emulative US decisions. In 1980, the Canadian commercial fleet harvested at a rate of about 40 kilotons per day and required just over two months to harvest approximately 2600 kilotons of halibut. Only a decade later, the fleet was harvesting at a rate of 644 kilotons per day and required only six days to exhaust the quota of 3900 kilotons. In that decade, the capacity and physical productivity of the fleet had increased by 30 percent per annum to the point that almost fifty percent more halibut could be harvested in one-tenth the time. A statistical analysis by Desharnais (*op. cit.* p.31-32) indicates that the policies of the IPHC, by increasing stock density, induced a more rapid entry of vessels and hence of fleet harvesting capacity. Canadian halibut vessel owners approached Fisheries and Oceans Canada for assistance in developing an individual quota (IQ) program. This led to an Individual Vessel Quota (IVQ) system which was implemented experimentally in the halibut fishery in 1991, after extensive industry input. After two years, an evaluation of the experiment indicated that the Department's conservation and management goals were being met. Additionally, a survey of vessel owners, indicated more than ninety percent in favour of continuing the IVQ system. Although modified in detail since then, the essentials remain of their rights-based system.

USA

As noted earlier, the US fishery is dominated by the Alaskan catch. Alaska's waters sustained abundant halibut fishing for local fishermen for a century and for indigenous people since time immemorial. By the 1970s and 1980s, that sustained abundance had begun to change dramatically. The local fishery had grown in size and fishing power and was extracting ever larger catches from the sea. By the 1990s, the Alaskan fishery was in serious decline. Fishery managers resorted to aggregate catch limits in an attempt to conserve the resource. In biological theory, such an approach can conserve the resource. However, as in the Canadian fishery, aggregate catch limits did nothing to stop the race to fish but did induce a downward spiral in the length of season as regulators and fishers tried to outpace each other. In such a regulatory regime, fishers, whose economic lives are endangered by the ever shorter seasons generate intense political pressure to delay and emasculate the limits. As a result, the limits tend to be too timid and too late to stop the fall. Given fewer fish and less time in which to catch enough to survive, each vessel and crew do all in their power to maximize catch before the season closes. The result is to further shorten the season.

1995 is the end of period for this section because in that year, Alaska adopted a rights-based system called Individual Fishing Quotas (IFQs). Because the Canadian halibut fishers had already had their IVQs in place for a few years, word had spread about some of the effects of the IVQ program. Also, in a timely survey Berman and Leask (1994) reported the results of their Alaskan survey of potential IFQ holders to determine their expectations for the incipient regime change to IFQs. Although the respondents surely knew some of the results of the IVQ program in Canada, it is unlikely that everyone knew (or believed) in them. In any case, we defer a discussion of the effects of the Canadian IVQ program to a post-1995 discussion. The ISER survey findings (n = 391; sampling error on percentages reported 4 percent), are interesting and we report them seriatim:

Fairness and Safety

Figure 1 summarizes the opinions of captains about fairness and safety effects of the incipient IFQ regime. Respondents were asked whether the IFQ system would be fair and improve safety.

Figure 1: Boat captains' Opinions

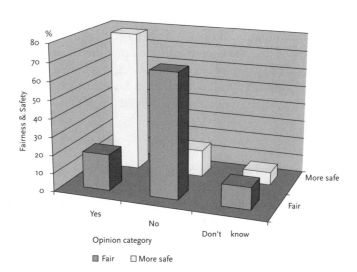

As shown in the bar graphs of Figure 1, about two-thirds believed that IFQs would not be fair. On the safety issue, more than three-quarters believed they would improve safety. In the words of one respondent:

> ".. IFQs will create a whole new set of problems but they will take away some of the danger".

Another longliner is quoted as follows:

> ".. IFQs may be the best of worst worlds... I don't think a group of people should own all the resources in the ocean".

Boat Captains Expectations for receiving IFQ Shares

Figure 2 shows the percent of captains expecting to receive IFQ shares. For this writer, the expectations are remarkably uniform; especially in light of the unfairness perception of Figure 1. Fairness appears to have had dimensions other than receiving IFQs. An obvious missing dimension is the fraction received. While only fifty-seven percent of the large vessels expected to receive IFQ shares, the graph says nothing about size of share.

Boat Captain's Expectations and Preferences for Management Regimes

Figure 3 summarizes opinions of captains about management alternatives discussed at the time. These included IFQs, the status quo, and limited entry. The survey also included two residual categories "Don't Know" and "Something Else". The Figure indicates that the IFQ option commanded the largest percentage of support among the realistic candidates. It also indicates that fully forty-two percent

Figure 2:   Captains' Expectations

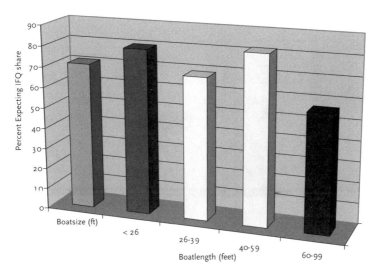

either were unsure (don't know), or preferred something else. A total of thirty-six percent preferred either the Status Quo or Limited Entry options. It seems a reasonable summary to say that, while there was much dissatisfaction with the status quo, the support for the IFQ option was significant but not overwhelming.

Boat Captain's Expectation for the Effect of IFQs on their own Finances
Figure 4 summarizes opinions of captains about management alternatives and their expected effects on finances of the respondents. This Figure may help to explain the dispersion of Figure 3. Fully forty-two percent expected IFQs to worsen their finances. This is quite remarkable and indicates that the economic case for IFQs had not persuaded many. Given such scepticism toward what has been a major point in favor (among economists, at least), of such management approaches, the dispersion in Figure 3 is less surprising.

The ISER report contains widely known information, that prices to Canadian halibut fishers had increased substantially over Alaskan prices after the IVQ program was introduced in Canada. Given the security of use rights afforded by IVQs, fishers could choose when to harvest their limited catch. They quickly realized that fresh prices were more profitable. We will return to this price differential later. It seems likely that the driving influence behind the opinions of captains were safety considerations and possibly price expectations. I find it hard to argue that expected financial returns were a dominant factor. Of course, there is the classic example of economic surveys which suggests that people will seize on any explanation to avoid the appearance of ignoble motives[10].

---

10   To apply this formula, one needs an estimated demand function, to constrain the values of Dp and DQ.

Figure 3: Captains' Preferences for management Options

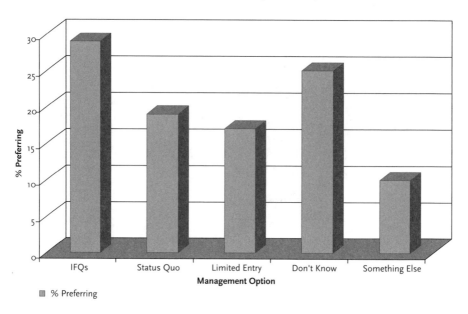

Figure 4: Expected Effects of IFQs on your Finances

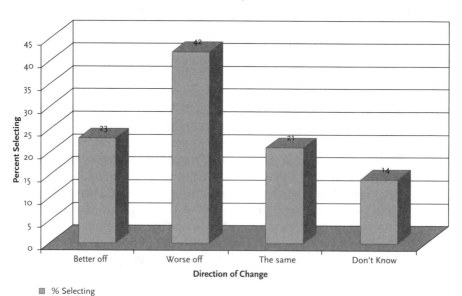

Figure 5: Canadian Fishery: Landed Weights, Values & Prices

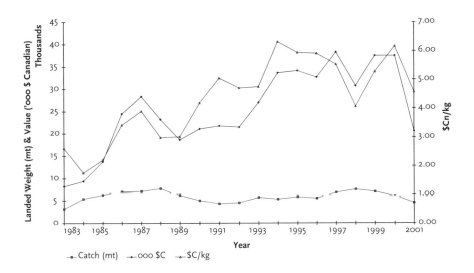

Retrospective: Post 1995 Events

In this section I report what has happened since introduction of the Alaskan IFQ program and the Canadian IVQ program which actually predates 1995. A slight anachronism seems an acceptable price in return for inclusion of both in the retrospective review of this section. I conclude this section with some quotes from fishers who have expressed their opinion on the new reqime, since the opinions of fishermen must count heavily in evaluating its success or failure.

Status of the stocks

Figure 5 contains data on the Canadian fishery. There are two noteworthy things about this figure; (1) the stability of landings and (2), the near tripling of ex-vessel prices over the period.

There is no obvious increase in landings that would indicate stock recovery following introduction of the IVQ system. That is to be expected since the Canadian and Alaskan fisheries share a common fishery but the Canadian fishery is only one-tenth as large. We will have to examine the larger picture after the Alaskan IFQ if we seek indicators of stock recovery. If no stock recovery could be expected from the Canadian IVQ system, what was the attraction for the industry coalition which broached the topic with Canada Department of Fisheries and Oceans? We lack the equivalent of the ISER study before the IVQ implementation, but it seems likely that safety considerations played a large part. The price increases in this figure are consistent with our earlier discussion of markets in space, form and time and may also have been anticipated. Earlier we discussed an ISER survey of Alaskan fishers which was conducted after the Canadian IVQ program began but

Figure 6: U.S. Commercial Catch & Trends, AK & Total

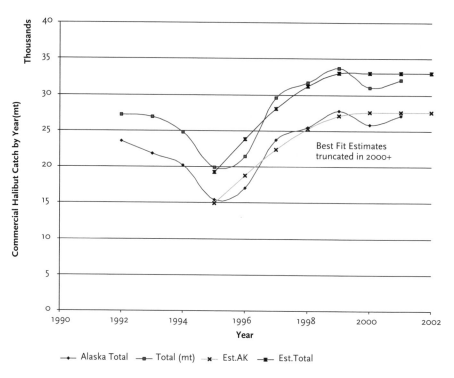

before the US/Alaskan IFQ program. The results seemed to indicate that safety at sea was a major determinant of attitudes toward the incipient IFQ program. We also have a more recent ISER survey that bears on this question (Knapp, 1999a). Although the findings are anticlimactic at this point, it is certainly worth stating them. The overwhelming majority (89 percent of Resurveyed Captains and 85 percent of Permit Holders), felt that IFQs had made fishing safer. Among those who also felt that their finances had improved, 95 percent felt that IFQs had made fishing safer. Interestingly, among those who felt their finances had deteriorated under IFQs, 14 percent felt that IFQs had made fishing less safe. I think it is chance that this is exactly the same percentage who selected the ("Worse finances", "a lot" (more discards)) combination, as discussed above. It is likely, in my opinion, that the ISER survey respondents allowed the financial outcome to contaminate their opinions on other aspects of the survey; most notably the discard and safety questions. Nor is it clear, on the discard question, whether they are describing their personal actions or the actions they ascribe to others. This is a common enough problem in survey research. On the other hand, on the safety issue, fully 74 percent of those who selected the "Worse off " (financially) option, percent felt that IFQs had improved safety.

Figure 6 contains data on Alaskan and total US landings since 1990, plus best fit curves to the data since 1995. The best fit curves should not be taken too seri-

Figure 7:   Landed Weight; Standardized by State to Force Commensurate Scales

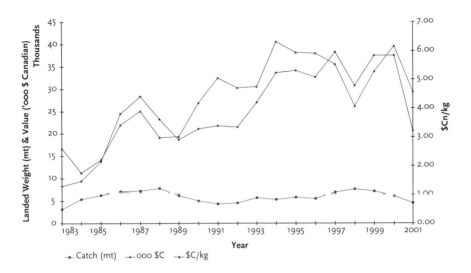

ously as forecasts; in fact, I truncated them for post 2000 since the functional form used would have had landings decreasing after 2000. Forecasts of stocks and landings of a single species are notoriously unreliable because of many interactions and random events, such as the El Nino-Southern Oscillation.

It is evident that the 1990-1995 period was one of declining supplies which contributed to the price increases also experienced by Canadians before introduction of their IVC program. The US fishery harvest experienced an annual rate of growth of 8.3 percent during 1995-2001; a trend which may or may not have been caused by IFQs. It can be objected that the observed recovery may be a coincidence. Maybe. The same can be said also of recovery following any management regime. One Robin does not make a spring. A check of the IPHC Newsletter for March, 2002, includes the following statement: "For the past five years, the IPHC has applied a harvest rate of 20 percent, down from a rate of 35 percent used in 1985"

It seems reasonable to me to infer that this increased conservation was assisted by the new regulatory regime begun in 1995. The IPHC Newsletter goes on to project a decline beginning in 2005 due to an oceanic regime shift which began in 1998-1999 and which is expected to begin depressing halibut recruitment in 2005.

Figure 7 displays data on weekly landings by State and year during 1990-2000. The descriptive statistics for weekly data were calculated and then converted to standardized Z variables to force similar scales on the data and to detect patterns in central tendency and dispersion. The near coincidence of Alaskan and total landings is quite evident. The trends by state indicate a drift in landings toward Alaska. California's standardized landings actually increased during 1994 -1996, then swiftly declined toward 1990 levels.

The corresponding trend for Oregon shows a steady decline from +1 standard deviations in 1990 to -1 standard deviations in 2000.

Figure 8:   Percentage Monthly Halibut Landings 1995-1998

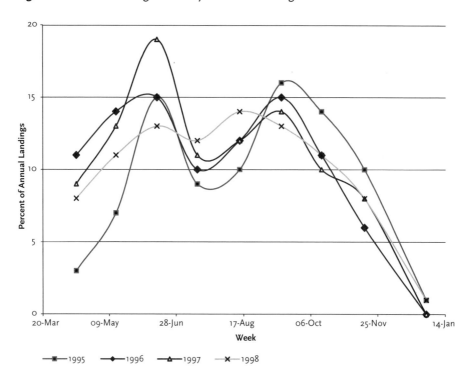

If the pre-1995 regulatory regime was causing a race to fish and a less uniform seasonal distribution of landings, this should appear in the post 1995 landings data as a more uniform seasonal pattern. The monthly data were expressed as a percent of the annual totals and displayed by year in Figure 8.

Bear in mind that markets do not adjust instantaneously to large structural changes. With this in mind, Figure 8 suggests that the new regime is generating a more uniform seasonal distribution of landings[11]. Continuing this thought, let us examine some statistical measures of dispersion. The coefficient of variation (CV) expresses variability (the standard deviation) relative to the mean. The kurtosis statistic is a measure of the likelihood of extreme events; the Normal distribution has zero kurtosis. A positive statistic indicates "fat tails" or platykurtic distribution. The skewness is a measure of symmetry about the mean; many economic variables have positive skew. Dispersion measures are important to markets In the present setting we are interested not so much in their general level as in whether they show any patterns of change during the period in which markets were adjusting to the new possibilities opened by IFQs. Figure 9 summarize these statistics. There is no obvious trend in the coefficient of variation, which was nearly constant. However, a

---

11   http://www.nfcc-fisheries.org/ir_pov_c09.html

Figure 9:   Monthly Harvests-Measures of dispersion

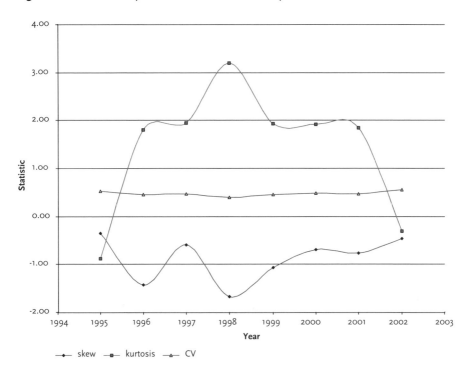

constant coefficient of variation, when the mean is increasing (as it did since 1995), also implies increasing variance since 1995.

The skewness varied somewhat but there is not a strong pattern. However, the kurtosis statistic indicates a strong increase during 1996-2001 (peaking in 1998), in the occurrence of extreme seasonal landings. This was a period of rapid change, inventory accumulation as expectations were not realized, followed by price discounts and inventory liquidation. Again, this sequence is not surprising given markets' aversion to risk (extreme events), and the enormous structural changes made possible by IFQs. Buyers do not much mind seasonal fluctuations in supply because they are predictable and are anticipated. However, fluctuations in landings which are larger than anticipated, require rapid renegotiation of fresh sales volumes and/or increased change of form to frozen product. Recall our earlier discussion of markets in space, form and time and the quote from Crutchfield and Zellner's 1962 study on the importance of smooth, predictable flows of raw product.

Price Trends

The IVQ program enabled Canadian halibut fishers to optimize time of harvest and product form. But a part of the price trend in this figure is due to supply decreases from Alaska; we also need to consider the larger fishery. If the upward price trend is purely a matter of aggregate supply, Alaskan and Canadian prices would have

moved in lockstep with each other. In fact, the ISER survey noted a widening gap between Canadian and Alaskan prices in favor of the former after the IVQ system was introduced in Canada.It is probably impossible at this time to fully explain all the changes that have occurred in the market for halibut. We have the following structural changes and shocks to the system during the 1990-2000 decade:
- multiple supply points over a 3000 mile range
- a multi-nodal transport network with fresh and frozen product forms.
- product form price differentials
- Canadian adoption of an IVQ system in 1991
- attendant changes in length of harvest season and product mix
- Alaskan adoption of IFQ system in 1995
- season length increase from 2 days to 245 days
- increased fresh product mix in 1991 and 1995
- expectational errors resulting in large inventory fluctuations during 1997-1998.

There have been several demand studies over the past decade (published in the Marine Resource Economics journal), which have attempted to capture the effects of structural change in this fishery as they unfolded. However, the data available has been incomplete and there has been a sequence of shocks. Some shocks, like the fresh product mix changes from British Columbia (1991) and Alaska (1995) are superficially similar. Yet, while an increase in fresh product from 10 percent of suppliers may be welcomed by buyers and sellers, an increase by the other 90% must have been a time of both great risk and great opportunity. The former increase may be absorbed easily by existing market channels. The latter may require development of new channels which takes time. Although they are among the most powerful quantitative tools available to economists, statistical analyses of time series market data is ill-suited to forecasting structural change. Estimation of the effects of structural change works best after an equilibrium pattern has been re-established. I present below an intuitive statement using graphs and measures of dispersion on the assumption that markets are risk averse and respond cautiously to structural changes. It is to be hoped that more definitive econometric estimates are forthcoming. Moreover, higher prices to fishermen are not an entirely appropriate objective for evaluation of program benefits, because the highest prices are obtained under a monopoly. Larger supplies of lower cost fish mean lower prices to fishermen but benefits to consumers that can be measured by changes in consumers' surplus. The standard measure is ΔCS:[12]:

$$\Delta CS = {}^1\!/_2\, \Delta p \Delta Q$$

where: ΔQ = quantity change
Δp = absolute value of price change induced by ΔQ

---

[12] To apply this formula, one needs an estimated demand function, to constain the values of Δp and Δq.

Figure 10: Pacific Halibut Price Data - US, Canadian/US Price Ratio

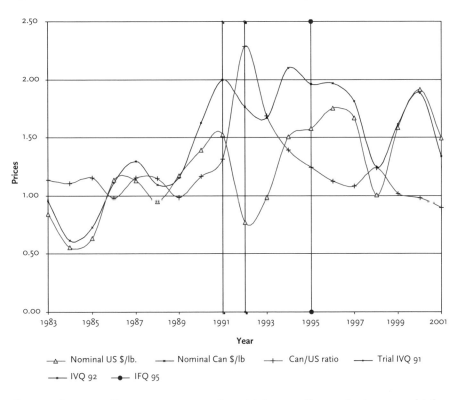

Conversely, upgrading a constant supply to higher quality products means higher prices to consumers which can also be measured as a benefit to consumers using a more complex measure for quality-induced changes in consumers' surplus.

Figure 10 contains the mean US and Canadian nominal ex-vessel price data (converted to US dollars per pound, and the ratio of the two, from 1992-2000.

The US average annual price series indicates a dip in 1992 when Canadian IFQs were inaugurated, and again in 1998. A drop in 1991-1992 US prices may have been due to increased fresh product competition from Canadian landings under their IVQ program which was phased in during 1991-1992. It is reported that Canadian share of the higher priced fresh market increased during this period.1998 was a disastrous year for prices. A possible explanation may lie in the large increases in landings as stocks rebounded after IFQs were adopted in 1995 and inventory carryover from the preceding year had to be sold off. The standard deviation series on prices (Figure 10), indicates decreasing month-to month price variability after 1992 which may be due to loss of fresh market to Canadian competition. From 1995 through 1998, price variability, as measured by standard deviation, remained at very low levels. Since 1998, variability has increased again which may be due to recapture of some fresh market share. The increased level of price variability is still less than the pre-1992 levels. To put the matter of fresh markets

Figure 11: Seasonal Price Variability Statistics-US

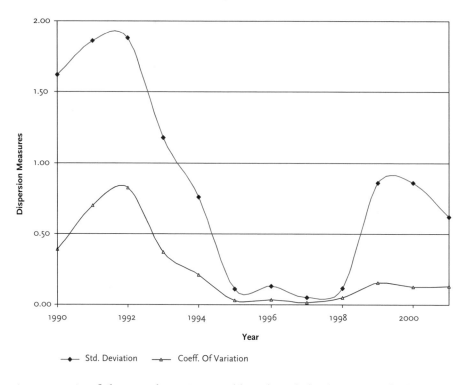

in perspective, failure to adopt IFQs would not have helped recapture fresh market share from Canadian suppliers; it would have doomed it to failure in advance.

The advantages of rights-based fishing for rational input combinations, harvesting and conservation decisions are discussed elsewhere in this paper. However, there can also be interactions with product quality. To see why this may be so, think for a moment of the decision alternatives facing a captain at sea. On the one hand, he can opt for return so as to preserve high quality and price. Conversely, he can stay at sea for another day to harvest a few more thousand pounds of fish. There is a stark trade-off between quantity and quality/price. In New England, numerous, well-intentioned efforts have been made to persuade fishers to adopt "on-board boxing" of fish. When experiments were done, the practice is a marginal proposition. Part of the real cost of the process is the time of crewmen not fishing. In the absence of exclusive, output-based harvest rights, the race to fish can trump marginal quality considerations. Because it is irreversible, quality deterioration at sea carries over on land whether the product form is fresh (higher priced) or frozen (lower priced). The identification of high priced with fresh product is confounded by the fact that it is perishable while frozen product is durable. At a sufficiently low price in makes economic sense to divert fish into frozen form and hold it for future demands. Unfortunately, I am not aware at this time of any empirical study that measures the separate effects of rights-based

fishing on raw product quantity, quality and product form. The Wilens and Homans (1994) paper lays out the marketing issues, and their importance for fisheries management but more work is needed on this matter. Certain facts are known, however.
- After introduction of IVQs in British Columbia, the fraction of their fish entering the fresh fish product form increased from 40 percent to 94 percent (Herrmann,1999, Appendix 3).
- Alaskan fishers did not enjoy a similar increased participation rate in the fresh market immediately after IFQs were introduced.

If we conjoin the earlier remarks about markets in space, form and time, and the quote from Crutchfield and Zellner, it seems likely that space/transport costs may account for the disparate experiences of Canadian and Alaskan fishers in the immediate post IVQ period. A similar hypothesis is offered by Knapp (1997). Examination of Figure 9 suggests that the Canadian:US price ratio had returned to historic levels by the late 1990s.

Input Stuffing and the Race to Fish

Previously, we saw a general description of management regimes over the last century, culminating in dramatically shortened seasons. By 1995, the official season had been reduced to two days. What had once been an annual fishery had been reduced to a classic "derby" fishery. Crews worked non-stop for 48 hours without due regard for dangerous sea conditions and overloaded vessels. The season's tally, duly reported in the media would include vessels and lives lost due to the dangerous operating conditions and the need to survive economically in the regulatory regime forced on the vessels. With limited time, each minute became precious; crews would cut entangled long lines, rather than disentangle them, and the lines would drift, continuing to "ghost fish" for whatever species swallowed the bait and hooks. Gentler handling and release of undersized halibut became an unaffordable use of scarce crew time so such fish were torn apart and discarded with attendant low survival rate[13]. Such by catch is a frequent part of regulated fisheries but the problem escalates as time becomes dearer.

The topic of discards arises often in discussions of "quota-based fishing" such as that used in the IVQ and IFQ fisheries for Pacific Halibut. It is interesting, therefore, that fishermen's comments quoted above are on the "race to fish" discarding which took place under the old regime. Clearly, there are discards under both systems so the matter deserves further discussion. Fortunately, we have the benefit of ISER surveys on this question (Knapp, 1999c). The telephone survey was conducted in 1998 and the questions were for the year 1997. There were two populations surveyed;

---

13 Although sometimes presented indignantly, as waste of fish and evidence of moral failure, accelerated discards are also an example of "input-stuffing"; more labor is devoted to "productive" capture and less to "unproductive" disentanglement time. I know of no quantitative estimates to date on the importance of ghost fishing and discards induced by the race to fish.

- Captains fishing in 1997 who were also surveyed in 1993 ("Captains' Resurvey"), and
- a survey of 1997 permit holders ("Permit Holders Survey").

The sample size and response rates for the Captains' Resurvey were 249 and 84 percent, respectively. For the Permit holders, the analogous numbers were 200 and 67 percent respectively. Each respondent was asked,

> *"How much do you think was caught and then discarded without being sold or reported-very little, some or a lot?"*

A plurality of respondents chose "very little", with the percentage being slightly lower (but not significantly so) among Resurveyed Captains. Only 8 percent of Permit Holders and 14 percent of Resurveyed Captains selected the "a lot" response.
- The survey also asked respondents if they were
- "Better off financially with IFQs",
- "About the same financially" or
- "Worse off financially with IFQS".

Overall, 82 respondents selected "Better", 53 selected "About the same" and 155 selected "Worse". Knapp also made cross comparisons between financial and discard opinions. He found that the two sets of opinions were related; specifically, among those who chose the "Better off financially" option, 66 percent chose the "Very little" (discards) response. Conversely, among those who chose the "Worse off financially" option, 19 percent chose the "A lot" (discards) response. Knapp concluded that the respondent's financial success had biased their responses to the discards question.

The marketing outcomes of reduced seasons were devastating also. What had been annual landings were landed at the end of a single fleet trip. The theoretical pathology described in the discussion of markets in space, form and time, had become fact. A superior fresh product was now "forced" by a regulatory regime to be harvested at high cost and stored for an entire season. The product mix was necessarily tilted toward an inferior frozen frozen form and stored at high cost to serve in future markets. In a more sane regulatory framework, a larger fraction of the same harvests would be left in the ocean until needed to meet the demand for high quality fresh fish.

The survey results of Burnes (*op. cit.*, appendix), include the following comment by a fisherman[14]:

---

14 I am indebted to Burnes for a copy of her paper.

"... I love this program. I fish when I want, on a calm ocean, in daylight hours only. The income is predetermined and certain, the fishing is civilized and safe. I used to lose a lot of gear and leave a lot of dead halibut on the bottom. That problem is virtually eliminated, with short sets and no gear conflict. We used to shake all our rockfish (dead) as we had no time or room for them. Now I land all of them."

Again, on p. 20, Burnes explains the post IFQ redesign of gear as follows:

".. New gear focuses on durability, rather than speed (the rate at which hooks can be baited and set)."

These statements are as clear as one could ask for on the race to fish, as well as the discard problem and it is clear that the IFQ program allowed fishermen to solve both. Burnes also quotes Knapp (1996) as indicating a decline in harvesting costs from $0.83 per pound in 1993 to $0.72 per pound in 1995.

Management Costs and Cost Recovery

Public Choice Theory (PCT) is somewhat ambiguous about net changes in management costs in the post 1995 context. On the one hand, PCT predicts rent seeking by a bureaucracy would tend to raise costs. However, the IFQ program includes provisions for cost recovery from beneficiaries. PCT predicts that cost recovery engages the interest of fishers in ensuring the cost effectiveness of administration and enforcement. The budget for NMFS' Restricted Access Management (RAM) program was $1.2 million (net of NMFS charge of five percent). However, expenditures by RAM have fallen since program inception. NMFS budget for enforcement in the halibut fishery increased from about $2.75 million in 1995 and 1996, to $3.75m in 1997. These amounts are less than what NMFS requested. Given the short time elapsed it would be rash to forecast the final outcome. However, it looks like both parts of the PCT predictions are consistent with events. For details on the cost recovery program, see NMFS (2002).

Processing Costs

Knapp and Hull (1996) reported increased costs in the processing sector due to reporting costs (ten percent higher) and higher labor costs (also ten percent higher). The higher labor costs seems anomalous since a more uniform seasonality would be expected to lower labor costs.

Consolidation of IFQ Shares

Of all the controversies about rights-based fishing, few are more controversial than the fear of consolidation. Opponents fear that large corporations have the wherewithal to pay more for shares and will therefore acquire undue amounts of a resource which local fishers had come to regard as "their" fish. A coherent explanation for why we should expect this outcome is rarely if ever given. Although no legal right existed before, Ciriacy-Wantrup's concept of "use-rights" are well estab-

lished among fishers which leads to strenuous objections to any change infringing those use rights, however extralegal they may be. Economic theory is ambiguous on this controversial topic of consolidation. Among the considerations are the following:
- Economies of size may put very small operators at a disadvantage, but this can also be true under other regimes as well.
- Most studies of vessel economics do not show much economies from very large vessels or multiple vessels. The size of vessel is dictated largely by the operating environment; offshore vessels must be large to be sufficiently seaworthy
- There may be economies of risk reduction from holding shares across fisheries for which revenues have negative covariances
- Introduction of rights-based approaches often includes safeguards against undue consolidation.

In the halibut IFQ fishery there were restrictions on the maximum percentages any individual could hold. With some technical exceptions, this was one percent. The Alaskan Limited Entry Commission contains a detailed report on consolidation as of 1998. Here is a summary of the findings.

The maximum percentage of shares which any person or corporation could hold varied by region but was in all cases a small percentage (around 1.5 percent or less). In the more remote western areas, there were "set-asides" for Community Development Quotas (a form of collective access right allocated to a community). In some cases this required compensation of individuals or companies who would have had (more) Quota Shares based on fishing history. The cost of compensation was spread uniformly over all IFQ holders. Subsequent transfers have been documented and tabulated in a report available on the website of the Alaskan Limited Entry Commission[15]. In some cases, the information on which initial allocations were based proved inaccurate and resulted in revocation of some quota shares. The number of persons holding shares declined as a result but in no area did the number of disqualified persons exceed 30. Also, some persons receiving quota shares were small and part time lumbermen/fishermen, and subsequently sold shares. As a result, the category in which the greatest decline occurred was persons who held less than 0.25% of shares in a region. The declines varied by region from a maximum decrease of 39% decrease to a minimum decrease of 0%. In larger classes (in terms of percentage holdings), the direction of change varied; in some areas larger holders increased their share; in others their share decreased. Despite the decrease in numbers of holders with < 0.25%, the total shares increased in this category also. Overall, quota shares increased as the stock rebounded and more shares were issued. An examination of changes in shares by vessel class shows that, while most vessel categories increased their quota shares, the largest percentage increases were for the smaller vessel classes (60 feet and under). The

---

15 http://www.cfec.state.ak.us/research/

freezer vessels received modest increases and even decreased in some areas. In general, the median holdings remain substantially less than average holdings indicating that they are skewed toward small holders. It appears that the IFQ program design and implementation has thus far successfully avoided a priori fears of many that IFQs would cause excessive consolidation of quota shares.

Testimony of Alaskan Fishermen on the IFQ Program
Arne Fuglvog is president of Petersburg Vessel Owners, and a member of the advisory panel to the North Pacific Fisheries Management Panel, the IFQ implementation team and the International Pacific Halibut Commission Research Advisory Board.[16]

> *A lifelong Alaskan, I have been commercially long lining since 1975. I was a deckhand during the Individual Fishing Quota (IFQ) qualifying years and therefore did not receive any quota shares at initial issuance. I have since purchased quota and bought into the fishery, as well as being a hired skipper. In my opinion, the Alaskan Halibut and Sablefish IFQ program has been a success to this point. There are more positive benefits then negative effects. As a fisherman in these two fisheries, I believe that the greatest benefits of the program have been: Conservation of the resource, safety at sea, improved quality, higher ex-vessel prices, and higher catch-per-unit-effort.*

Conservation of the Resource
Reduced halibut and sablefish bycatch and discards, changing from the derby system to IFQs allowed for the retention of both species when IFQ fishing.

- Reduced fishing mortality from lost gear
  Again, changing from the derby to an IFQ fishery has lowered the estimates of halibut mortality due to lost gear by almost a million pounds of halibut a year.
- Reduced bycatch and discards of non-target species
  Slowing the fishery down has allowed fishermen to retain and properly handle non-target bycatch species. This has increased the quality and value of these species.
- Staying under the Total Allowable Catch (TAC)
  Before 1995, both the halibut and sablefish TAC was exceeded in a number of seasons. Since 1995, neither the halibut nor sablefish TAC has been exceeded even once, in any regulatory area.

---

16 Arne Fuglvog 'The Positive Benefits of the Halibut and Sablefish IFQ Program', http://www.nfcc-fisheries.org/ir_pov_c09.html

Safety At Sea
Having the flexibility to choose the time of year, weather, and area to fish, has decreased the loss of vessels and lives, as well as search and rescue cases.

Improved Quality
Switching to an 8-month season has greatly increased the quality of catch, especially for halibut. Now, more than 80% of the halibut goes into the fresh market.

Higher Ex-Vessel Prices
Ex-vessel prices are higher since 1995. This year, as an example, halibut has maintained a price of around $2.50 per pound, and sablefish was more than $4.00 per pound this spring.

Higher Catch-Per-Unit-Effort (CPUE)
Sablefish CPUE increased approximately 30% after IFQs were implemented. Increased CPUE has meant decreased expenses such as fuel, bait, gear, and groceries, relative to inflation.

Clearly, not all of the effects of the Halibut and Sablefish IFQ Program were positive. I will briefly list a few of the negative effects that I have observed:
- Net loss of crew jobs,
- Initial investment, and
- Expense of purchasing quota.

Net Loss of Crew Jobs
There has been a net loss of crew jobs. Even though many deckhands have bought into the program, overall there are fewer jobs. Many of the crewmen fished off vessels that received substantial amounts of IFQs, but reduced crew after implementation.

Initial Investment
Because of the high costs of purchasing IFQs, it takes so much more of an initial investment now to get into the fishery. This cost is spread out by having fewer crewmen or by having crew shares reduced. Both are negative effects.

Expense of Purchasing Quota
Many individuals received small amounts of quota and have sold out. Much of the quota that was purchased early on was by large initial recipients, who had the capital and the collateral. It is still difficult for an initial entrant to find the right quota to purchase, come up with the down-payment, have sufficient collateral and income, and receive financing. The National Marine Fisheries Service's new loan program will help, but many of the small villages in Alaska feel disadvantaged in this process.

I believe that Congress should lift the moratorium on the development of new IFQ programs. I support the National Research Council's recommendation to Congress, that IFQs may not be appropriate in all regions for all fisheries, but they are a useful tool that the regional councils should have at their disposal. There are a number of fisheries that could benefit from an IFQ system. If the moratorium is not lifted, we won't be able to analyze whether an IFQ program would be the best way to rationalize these fisheries.

Summary, Conclusions and Caveats

The Pacific Halibut fishery has a long and well documented history which is an envelope to the various ideas of fisheries management in the 20th century. In the closing years of the century, the Canadian and then the US fishery switched from aggregate quotas and closed seasons to rights based fishing. The experiences of both fisheries with the new regime have been overwhelmingly positive. Major positive benefits have been:
- Improved safety and less loss of life
- Improved resource conservation
- Higher landings
- Lower harvest costs
- Higher quality product
- Higher prices to fishermen
- Higher consumer welfare
- Less discards
- Less ghost fishing
- Lower costs for product storage and distribution
- Creation of valuable assets for the retirement portfolio of quota holders

Among the negative attributes (listed by some fishers) are
- Loss of some jobs for crewmen
- Higher cost of entry

The negative items listed overlap and are simply another side of the same coin. For example, the cost of quota is a forced savings and retirement program, as is the cost of any business asset to an aspiring small business. The regime change provides the income flow to pay into this retirement program. In the absence of rights based fishing, any fisherman can contribute to a retirement plan of his choosing. Unfortunately, the associated income stream is likely to be too small to allow him to do so. Similarly, crew could have been allocated shares but were not. This is a distributional issue for which there are human values but no uniquely correct answer. In the long run, however, society is better off if labour moves to jobs where the value added is at least as great as opportunity cost. This allocative principle was violated in the Pacific halibut fishery until rights-based fishing began a decade ago.

Caveats

Rights-based fishing is a collection of approaches of which the IFQ and IVQ systems are examples. It would be foolish to suggest that such systems are appropriate in all fisheries and I know of no one who makes such a sweeping claim. However, the conclusion seems inescapable that in the Pacific halibut fishery, theory and results are in good accord and rights based fishing has been a resounding success.

References

Bell, F.H. 1981: *The Pacific Halibut: the resource and the fishery.* (Anchorage: Alaska Northwest Publishing Company).

Berman, M. and L. Leask.1994: "On the eve of IFQs: fishing for Alaska's halibut and sablefish," *Alaska Review of Social and Economic Conditions* XXIX(2), November. University of Alaska, Anchorage.

Burkenroad, M.D. 1948: "Fluctuations in Abundance of Pacific Halibut," *Bulletin of the Bingham Oceanographic Collection,* May, 1948.

Burnes, E.I. 1998. "The political economy of the North Pacific halibut fishery under individual fishing quotas," California State University, Fresno.

Casey, K.E., C.M. Dewees, Bruce R. Turris and James E. Wilen. 1995: "The effects of individual vessel quotas in the British Columbia halibut fishery," *Marine Resource Economics* 10( ):211-230.

Crutchfield, J.A. and A. Zellner (eds.). 2002: *The Economics of Marine Resources and Conservation Policy: The Pacific Halibut Case Study with Commentary.* (Chicago: The University of Chicago Press). Reprint of 1962 classic.

Crutchfield, J.A. 1981: "*The Public Regulation of Commercial Fisheries in Canada Case No. 2, The Pacific Halibut Fishery,*" Technical Report No. 17, Ottawa, Ontario: Economic Council of Canada.

Crutchfield, J.A. 1961: "*Regulation of the Pacific Coast Halibut Fishery,*" Expert Meeting on the Economic Effects of Fishery Regulation, Ottawa, Canada, 12 to 17 June 1961 (May, 1961), "Panel IV Discussion," 7.

Crutchfield, J. and A. Zellner. 1962: "Economic Aspects of the Pacific Halibut Fishery," *Fishery Industrial Research,* 1(1): 102.

Desharnais, C. Undated: "Success and Failure Under Regulation 1930-1960", MA thesis in History, University of British Columbia, Vancouver, Canada.

Gates, J.M. 2000: "Input Substitution in a Trap Fishery," *ICES Journal of Marine Science,* 57: 89-108. Agr. Exp. Sta. Cont. No. 3672.

Gordon, H.S. 1954: "The economic theory of a common property resource: the fishery," *J. Political Economy* 62:124-144.

Herrmann, M. 1999: "The relationship between ex-vessel revenue and Halibut quota: some observations," Appendix 3 in report to the North Pacific Management Council, November 9, 1999.

Knapp, G. 1997: "Alaska halibut markets and the Alaska IFQ program," Final Report for the Saltonstall-Kennedy Program (NOAA Grant # NA371Ed0184). Anchorage, AK.

Knapp, G. 1999a. Effects of IFQ Management on Fishing Safety: Survey Responses of Alaska Halibut Fishermen. *ISER Working Paper Series: Surveys of Alaska Halibut Fishermen about Effects of IFQ Management,* May, 1999. Institute of Social and Economic Research, University of Alaska, Anchorage.

Knapp, G. 1999b: *Effects of IFQ Management of Resource Conservation: Survey Responses of Alaska Halibut Fishermen.* Institute of Social and Economic Research, University of Alaska, Anchorage.

Knapp, G. 1999c. "Unreported Discards Under IFQ Management: Survey Responses of Alaska Halibut Fishermen". *ISER Working Paper Series: Surveys of Alaska Halibut Fishermen about Effects of IFQ Management,* May, 1999 Institute of Social and Economic Research, University of Alaska, Anchorage.

Knapp, G, and D. Hull. 1996: The first year of the Alaska IFQ: a survey of halibut quota holders, Institute of Social and Economic Research, University of Alaska, Anchorage.

Laukkinen, M. 2002: "Regulatory objectives in the North Pacific halibut fishery: how far is the Regulator from the economist's ideal?". UBC ARE working paper, # 933. June 2002. Personal copy courtesy of the author.

Mollet, N. (ed.) 1986: *Fishery Access Control Programs Worldwide: Proceedings of the Workshop on Management Options for the North Pacific Longline Fisheries*, Alaska: Alaska Sea Grant, December, 1986.

NMFS. 2002: *Annual Report: IFQ Fee (Cost Recovery) Program, Pacific halibut and sablefish Individual Fishing Quota (IFQ) Program*. Juneau, AK, National Marine Fisheries Service.

Pautzke, C.G. and C.W. Oliver. 1997: "Development of the individual quota program for the sablefish and halibut longline fisheries off Alaska". *Report to the National Academy of Science*, Anchorage, Alaska.

Richards, H. and A. Gorman. 1986: "The Demise of the U.S. Halibut Fishery Moratorium: A Review of the Controversy, " In: Mollett (ed.),(1986).

Skud, B.E. 1977: "Regulations of the Pacific Halibut Fishery, 1924-1976," *Technical Report No. 15*. International Pacific Halibut Commission, Seattle, Washington.

Skud, B.E. 1978: "Factors Affecting Longline Catch and Effort.," *Scientific Report No. 64*. International Pacific Halibut Commission, Seattle, Washington.

Shotton, R. (ed.) 2001a: "Case studies on the allocation of transferable quota rights in fisheries," *FAO Fisheries Technical Paper 411*.

Shotton, R. 2001b: "Case studies on the effects of transferable fishing rights on fleet capacity and concentration of quota ownership," *FAO Fisheries Technical Paper 412*.

Wilen, J.E. and F.R. Homans, "What do regulators do? Dynamic behaviour of resource managers in the North Pacific Halibut Fishery," *Ecological Economics*, (March 1998), v. 24, no. 2-3, 289-298.

The Experience of the Mauritanian Fish Trading Company (SMCP) in the Management of the Fisheries Sector in Mauritania

Stephen Cunningham, Sid'El Moctar Ould Iyaye and Debbe Ould Sidi Zeine[1]

Introduction

The early 1970s were characterised by the concerns expressed by countries from the South over their deteriorating terms of trade and the need for improved export prices for their raw materials.

Mauritania, which had just created its own national currency and nationalised its mines, needed to acquire hard currency reserves and to ensure adequate fiscal receipts to consolidate the new currency and meet budgetary constraints associated with the increasing social pressure resulting from the rural exodus following successive years of drought.

The limit to the territorial sea was increased unilaterally to 12 miles in 1971, and then to 30 miles en 1972, which allowed the country to sell fishing licences to most of the fishing companies operating along its coast. In this way the country began to generate fiscal receipts on a quite substantial scale compared to the needs of the time. The Ministry of Fisheries and the Marine Economy (MPEM) was created in 1977 and efforts were made to organise and regulate the sector. A Law covering the Merchant Navy and Marine Fisheries entered into force on 23 January 1978.

With the new international law of the sea, Mauritania declared a 200-mile Exclusive Economic Zone (EEZ) at the end of 1982.

---

1 Respectively IDDRA; Fisheries Ministry, Mauritania; Nouakchott University, Mauritania. The views expressed in this document are those of the authors alone.

Development of Mauritanian fisheries policy

The New Fisheries Policy (NFP)
The NFP was adopted on 18 October 1979. It put an end to the direct licensing of foreign vessels. It replaced this with a mixed system involving vessel chartering within the context of bilateral fishing agreements, encouraging the development of joint ventures, the development of onshore activities and the employment of a quota of Mauritanian sailors on fishing vessels operating in the Mauritanian EEZ. The development of artisanal fishing was strongly encouraged, even as a national industrial fishing fleet emerged.

The first fishing agreements under this new policy appeared in 1980. The national fleet began to develop, generally through the taking-over of Spanish or Korean freezer trawlers, initially via leasing arrangements with options to buy. A major leap forward took place concerning the role of the sector in the national economy. Total exports, which were only 1,000 million ouguiyas (UM) in 1980 increased to 4,221 million UM in 1982 and 9,722 million UM in 1983.

Strategic Note on Fishery Sector Development (1994)
Associated with a period of structural adjustment, the strategic note changed fisheries policy in line with the liberalisation and market economy orientations adopted by the Government. The State withdrew from public enterprises and foreign exchange was liberalised. The SMCP (see further below) was partly privatised, with the State keeping 35% of the capital of the company. A fishery management regime based on access to the resource was introduced for the artisanal fishery in 1995 and for the industrial fishery in 1997.

Management and Development Strategy for the Fisheries Sector (June 1998).
This strategy was based, inter alia, on the following goals:
– Access to the resource should be in line with the allowable potential determined annually for each fishery. There was an immediate freeze on new licences for the octopus fishery. The overexploitation of this fishery had been regularly reported by the national research institute (then CNROP, now IMROP). The same conclusion was reached by two international working groups in 1993 and in 1998. The activities of the artisanal fleet were to be included in the effort level that was to be allowed;
– A study was to be undertaken to examine the feasibility of converting some national cephalopod trawlers to other fisheries with development potential;
– Enforcement was to be increased to ensure compliance with a 2-month closed season ("biological rest period"). All demersal fishing was to be stopped during the octopus recruitment period;
– Particular attention was to be paid to fisheries research as an essential decision-making tool;
– The MCS system was to be improved and developed, with increases in both human and financial resources.

Resource access systems

Two broad periods can be identified when considering systems used to control access to the resource:

First period: Fishing rights

Prior to 1995, fiscal arrangements with respect to resource access were based on the quantities caught and landed. Fishers were required to sell their fish through the SMCP, which collected on behalf of the Government an export tax (called a Fishing Right) that constituted the backbone of the system. This Fishing Right distinguished between cephalopods and demersal fish, and between products that were frozen at sea and those frozen onshore. The amount to be paid varied according to the kind of vessel, national or leased, foreign.

In summary the system was as follows

	% of Turnover
Statutory Minimum Tax (IMF)	2
"Statistical" Tax (met running costs of system)	3
Fishing Right (or Export Tax):	
– Cephalopods, frozen onboard;	11
– Demersal fish, frozen onboard;	8
– Cephalopods, frozen onshore;	6
– Demersal fish, frozen onshore	4

Second period: Access rights and territorial rights (since 1995)

The IMF was abandoned in 1995, whilst the Fishing Right was phased out by 1997. It was replaced by an Access Right calculated in terms of Ouguiyas per GRT. For the artisanal fishery, the so-called Territorial Right was introduced. In summary, the system was as follows:

	1995	1996	1997	1998	1999
IMF (% of turnover)	0	0	0	0	0
Fishing Right (% of turnover)	11	8	0	0	0
Statistical tax (in % of turnover)	3	3	3	3	3
Access right per GRT (industrial fishery):					
– Freezer trawler;	13.231	46.310	60.950	60.950	60.950
– Freezer vessel other than trawler;	13.231	27.786	37.983	37.983	37.983
– Wetfish trawler;	9.681	33.884	43.450	43.450	43.450
– Wetfish vessel other than trawler;	9.681	2.0331	27.792	27.792	27.792
Annual territorial right (artisanal fishery)					
– Vessel < 12 m;	12.000	24.000	24.000	24.000	24.000
– Vessel > 12 m	24.000	48.000	48.000	48.000	48.000

Mauritanian marketing policy for fishery sector products pre-SMCP

During the early years (1980 to 1982), freezing facilities were being constructed and exports of demersal species were made entirely by trawlers landing and trading through Las Palmas. The contribution to the Mauritanian economy, in terms of hard currency and budgetary receipts, was reduced by the anarchy and amateurishness of the operators and their tendency to under-declare their activities so as to reduce the need to repatriate foreign exchange and avoid customs duties.

Vessel owners were generally inexperienced and the tendency was to use the services of management agencies. These agencies would help to locate and purchase the vessel, and recruit the crew, and then assist generally for instance with supplies and repairs. Selling the catch through these agencies became the way of guaranteeing payments for the vessel, as well as payments to crew and for supplies. But the system was open to abuse and the offices opened in Las Palmas by the Mauritanian Central Bank and the Mauritanian Customs Authority proved insufficient for effective control of the situation.

As a result, by Decree N° 82-145 dated 12 November 1982, the Government made it a requirement for demersal species to be landed in and traded from Nouadhibou. This Decree entered into application on a general basis from February 1983. It had an immediate effect increasing exports of such species from 25,520 tonnes in 1982 to 55,340 tonnes in 1983, notwithstanding various difficulties due to the lack of infrastructure.

In fact, in order to ensure the success of its policy the Government undertook significant port infrastructure investment, including the establishment of ships chandlers, the creation of a second port handling company, providing attractive credit conditions, and building a temporary floating dock.

Notwithstanding these developments, the repatriation of foreign exchange was still somewhat haphazard because of the management agent system. In order to obtain security for their advances of supplies and for their services, these agencies used a system of dual credit notes. The vessel owner received a credit note from the agency for the value of his exported catch, the agency received a credit note from the vessel owner for the value of their supplies and services, the latter credit note being dated for encashment before the former. Whatever risks this system may have had from the private sector viewpoint, for the Government the risk was that the former credit note would be artificially lowered and the latter artificially increased so as to shift profits offshore.

To avoid this difficulty, the SMCP was created by Decree N° 84-130 dated 5 June 1984. This company was given the monopoly of selling (exporting) demersal species which were subject to the requirement to land in Nouadhibou. The creation this monopsony put an end to the system of dual credit notes and began a new era in the marketing of Mauritanian fish produce.

The marketing system implemented by the SMCP

The SMCP system went through a number of phases.

First phase: June 1984 – June 1987
In this first phase, which was of an experimental nature, the system was managed unilaterally by the SMCP, leading to persistent tension with the producers. Part of the problem was that the Decree creating the SMCP was ambiguous on the nature of the relationship between the company and fish producers.

On the one hand, if the SMCP is merely a service provider, selling on behalf of the producer, then the latter should be paid the real export price, less the costs of the operation.

On the other hand, if the SMCP is considered to be a company purchasing the product from the producer at an agreed price and which the company then sells on its own account and at its own risk, then the situation is very different.

This ambiguity was not helped by Statutory Instrument R-163 of 13 November 1984, applying the Decree, which simply said that the SMCP took delivery of the product.

During this first phase, the SMCP set the price at which it would purchase from producers and paid them once the produce had been sold and shipped, deducting storage charges, FOB costs and other costs. A marketing commission was established jointly with the producers but it had only a consultative role and producers were not always shown bids from fish buyers received by the company.

Second phase: 1987 – 1991
A marked development occurred during this second phase under pressure from the industry. Following negotiations between the Government and the sector, a new Statutory Instrument, R-123 of 30 June 1987, was adopted under which a joint price commission was established. The same day, Statutory Instrument R-122 set out the way in which prices were to be determined stating that it would be "on the basis of the international price, depending on the offers received by the SMCP and information on the current market situation and likely trends".

The same Instrument also established the structure of costs and fees that were payable to the SMCP. Henceforth, the company was considered to be buying on its own account and bearing all marketing risks. However, the producer had access to the offers received by the company and could, if he wished, try to find a buyer prepared to offer more for his produce. But he was still paid at the general price set for the relevant 10-day period, and did not have access to the final price for the transaction.

Third phase: 1991 -1993
Whilst the SMCP remained a public company, collaboration with producers was reinforced by the formal establishment (Statutory Instrument R-219 du 15 November 1990) of a consultative marketing commission to assist the Director

General. This collaboration was reiterated in Decree 91-100 of 8 July 1991 which set out for the first time a clear and coherent legal framework. The SMCP now became a service provider working on behalf of producers who were to be paid on the basis of the actual price of their product, less the various costs and fees. The producer remained owner of the product until it was shipped.

Statutory Instrument N° 355 du 22 July 1991, implementing the Decree, made it a requirement for the SMCP to transmit to the commission all bids received and all other information in its possession. Producers were able to present their produce to potential buyers and seek bids. Their representatives were able to accompany those of the SMCP on trade and marketing missions. This was a transitional phase prior to opening the capital of the SMCP to the private sector and fully associating producers in marketing.

Fourth phase: 1993 onwards

With Decrees 93-036 and 94-100, the company was privatised: around 65% of its capital was taken up by the operators in the national fishing sector and by national banks, and the SMCP became of private company.

With the entry of the private sector into the capital of the SMCP, marketing changed again. Decree 93-024 of 28 January 1993 confirmed that the SMCP was merely a service provider, acting in the name of and on behalf of the producer, who is fully associated with the final selling decision.

Marketing policy was taken over by the Board of Directors, considerably reducing the prerogatives of management. The centralised decision-making process and the control exercised over supply, which constituted the basis of the centralised marketing system, were called into question. Each producer was entitled to negotiate directly with his customer, who was often his technical partner. The SMCP then simply rubber-stamped the contracts between the two parties. Marketing became a vital concern that each group of producers sought to control.

In 1995, value-added produce based on demersal fish were removed from the SMCP monopoly.

Some difficulties with the marketing system

The main difficulties and weaknesses of the SMCP marketing system are discussed in this section.

Precariousness: The future of the SMCP has always been uncertain. As early as 1986, it was decided that the marketing system put in place by the company would only last for five years, at the end of which a decision would be made about the future of the company. Since the opening of its capital, there have been persistent rumours that its monopoly on the trading of frozen demersal products will be removed, as was the case for value-added products, with a view to encouraging processing. The company has always therefore been operating in a provisional environment and it has been handicapped by its inability to take a long term view in planning its activities.

Legal status of the company: In its initial phase, the status of ownership of the produce was a legal artifice – the SMCP had a monopoly to market produce that it did not own, whilst the producer remained the owner of produce over which it had no control. This situation was the source of all the conflicts between the SMCP and the producers.

Reduced market share: This is due to both internal and external factors. Internal factors include a concentration of fishing effort on octopus which led to significantly reduced market presence in Europe and Africa, the development of other sources of Mauritanian produce with the appearance of European and Chinese vessel fishing under licence and not required to land in Nouadhibou and selling into the same markets, and to a lesser extent the development of landings in Nouakchott that initially escaped the control of the SMCP. There has also been a marked fall in octopus landings. On the external side, there has been a considerable development in Moroccan octopus production, and there are signs of changing food preferences amongst Japanese consumers who represent the main market for SMCP octopus.

Duality: The current marketing system for frozen produce (cephalopods and demersals) which allows producers to negotiate export prices directly with their customers restricts the company's ability to influence the market in the sense of increased the value of Mauritanian produce. If the SMCP finds a customer prepared to offer a high price, it informs producers who then re-negotiate prices with their own customers. But the SMCP is then unable to meet the demands of its customers and its credibility is undermined. This problem has yet to be resolved and is weakening the SMCP's role in the system.

The role of the SMCP in Mauritanian fisheries management

Despite its limits and the difficulties that it has encountered, the SMCP has contributed positively to the management and development of the fishery sector and to its integration into the national economy.

The SMCP has contributed to the quest for improved export prices by providing some countervailing power to large importing companies. It has been an important instrument for the Government in collecting royalties from the fishery and in ensuring the repatriation of foreign exchange. It has also developed extensive databases concerning the fisheries which are of great use for the management of the fisheries.

The company has also provided support to fishing operators, especially smaller ones, which has freed them from the problems involved in marketing fish products, particularly for export, and allowed them to concentrate on the production side of their activity.

The role of the SMCP in recovering budgetary receipts for the State

Receipts from the Mauritanian fisheries sector represent some 20 to 25% of overall Central Government budgetary receipts, showing the huge importance of the sector to the national Treasury.

The following table shows the development of overall receipt and of the different categories over the period 1986 to 2000.

	1986	1987	1988	1989	1990	1991	1992	1993
Fishing Rights	2.188	2.111	2.267	2.346	1.751	1.895	2.249	3.003
Access Rights								
Fines		44	208	157	106	104	64	18	102
Other		779	677	706	706	635	740	772	1.656
NON FISHING AGREEMENT	3.011	2.996	3.130	3.158	2.490	2.699	3.039	4.761
Fishing Agreement	455	1.038	855	1.089	1.664	1.102	1.263	1.594
TOTAL	3.466	4.034	3.985	4.247	4.154	3.801	4.302	6.355
State Budget	13.691	15.681	16.684	17.834	19.159	19.792	21.559	27.360
Fishing Right Contribution %	15,98	13,46	13,59	13,15	9,14	9,57	10,43	10,98
Access Right Contribution %								
Non Fishing Agreement Contribution %	21,99	19,11	18,76	17,71	13,00	13,64	14,10	17,40
Fishing Agreement Contribution %	3,32	6,62	5,12	6,11	8,69	5,57	5,86	5,83
Total Contribution %	25,32	25,73	23,89	23,81	21,68	19,20	19,95	23,23

	1994	1995	1996	1997	1998	1999	2000
Fishing Rights	2.324	2.803					
Access Rights			469	1.876	2.797	1.713	1.595	1.395
Fines	133	210	345	290	500	550	550
Other	1.572	1.358	1.456	973	467	552	597
NON FISHING AGREEMENT	4.029	4.840	3.677	4.060	2.680	2.697	2.542
Fishing Agreement	1.791	2.776	6.409	7.901	8.462	9.154	9.958
TOTAL	5.820	7.616	10.086	11.961	11.142	11.851	12.500
State Budget	29.156	33.212	43.188	46.470	49.440	53.358	61.316
Fishing Right Contribution %	7,97	8,44	0,00	0,00	0,00	0,00	0,00
Access Right Contribution %			1,41	4,34	6,02	3,46	2,99	2,28
Non Fishing Agreement Contribution %	13,82	14,57	8,51	8,74	5,42	5,05	4,15
Fishing Agreement Contribution %	6,14	8,36	14,84	17,00	17,12	17,16	16,24
Total Contribution %	19,96	22,93	23,35	25,74	22,54	22,21	20,39

The analysis of this table enables a certain number of conclusions to be drawn:
1. The importance of the fishery sector to State budgetary receipts is clear, with the total contribution varying between 19.2% and 25.73%. The State depends for a fifth of its resources on this sector and on the exploitation of a fragile, but renewable, resource.
2. The non-fishing agreement contribution has declined substantially and monotonically falling from almost 22% of the total in 1986 to just over 4% in 2000. The replacement of fishing rights with access rights had been particularly disastrous. In 1995, the last full year of the old fishing right regime, the contribution of the national fishing sector was still almost 15% of the State budget.
3. At the same time, the dependence of the State on receipts from fishing agree-

ments has increased from just over 3% in 1986 to over 16% en 2000. The signing of a new agreement in 2001 will have increased this dependence yet further.
4. Two comments may be made on this situation. First, the Mauritanian State is clearly vulnerable to a change in strategy either by the EU or by itself in the area of fishing agreements. Were the agreement to end for whatever reason (as was the case with Morocco), Mauritania would find it very difficult to replace the lost revenue. This difficulty is related to the second point, which is that revenue generated under the old fishing right regime represented resource rent extraction whereas under fishing agreements such revenue is almost entirely a subsidy from the EU to its fleet. Under circumstances where less resource rent is being extracted and more subsidy is being paid, the economic theory of fisheries management clearly predicts that exploitation levels will increase. In the already overexploited octopus fishery this is exactly what has happened with the 2002 Working Group held at the National Research Institute (IMROP) concluding that overexploitation had worsened since the previous group (in 1998).

In both the fishing right and access right cases, the SMCP has played a key recovery role for the State:
1. During the first phase, where the fiscal system was based on fishing rights calculated as an export tax, the SMCP collected at the source all taxes and duties that were payable on behalf of the State. As the SMCP was directly involved in marketing, all payments due to the State were correctly calculated without under-reporting on the part of producers. The system was transparent and applied to the whole sector, including both industrial fishing and artisanal fishing;
2. During the second phase, with the change to access rights payable in advance by fishers, the SMCP has continued to play a determinant role, both in terms of recovering the payments for the State and in funding the system by providing credit to fishing companies, the financial situation of most of which did not allow them to make advance payments for access rights. The SMCP thus paid on their behalf and then gradually recovered the payment from their sales. Without the SMCP, most operators in the sector would not have been able to continue in activity, and the company has hence played a decisive, if unforeseen, role in the implementation of this new fiscal policy.

The importance of SMCP in recovering duties and taxes is shown in the following table covering the period from 1986 to 2001:

Year	Amount recovered for the State by SMCP (millions of UM)	Total receipts excluding fishing agreements (millions of UM)	Amount recovered by SMCP (% of total)
1986	1,029	3,011	34
1987	1,250	2,996	41
1988	1,502	3,130	47
1989	1,454	3,158	46
1990	1,054	2,490	42
1991	1,109	2,699	41
1992	1,634	3,039	53
1993	2,507	4,761	52
1994	2,712	4,029	67
1995	2,897	4,840	59
1996	2,058	3,677	55
1997	1,056	4,060	26
1998	971	2,680	36
1999	1,031	2,697	38
2000	1,311	2,542	51

Source: Annual Statistical Bulletin. SMCP

The role of the SMCP in repatriating foreign exchange from seafood exports

Since the creation of the national currency in 1973 and her withdrawal from the franc zone, Mauritania's foreign exchange needs have been met mostly by the export of two products: iron ore and seafood.

In the early years after independence, iron ore exports constituted the main source of foreign exchange for the country, but from the beginning of the 1980s, with the slowdown of the European steel industry, the main customer for Mauritanian iron ore, fish resources became the main source of foreign exchange. Seafood became the number one export in 1984 and has remained so ever since.

Contribution of fishery sector to exports (millions of UM)

	1993	1994	1995	1996	1997	1998	1999
Total Exports	51,109	50,710	64,810	69,995	60,743	65,900	78,200
Seafood Exports	28,251	25,733	35,065	36,434	28,059	26,100	32,400
Fish Sector as %	55.30%	50.70%	52.40%	52.10%	45.90%	39.6%	41.4%

Source: CEAMP et ONS

Despite changing foreign exchange policies in the context of structural adjustment programmes, the SMCP has continued to play a key role in ensuring the fluidity and regularity of hard currency repatriation.

In the phase immediately following the creation of the company, 70% of foreign exchange receipts had to be repatriated and paid into the Central Bank. The producer received the equivalent in local currency at an exchange rate set by the monetary authorities. The remaining 30% could be used to meet operating expenses incurred in foreign currency.

The centralisation of exports through the SMCP allowed foreign currency to be repatriated quickly, whilst avoiding under-reporting.

In a second phase, in a context of foreign exchange liberalisation, it was decided that 100% of foreign currency receipts could be held in a foreign-currency denominated account with the requirement that the producer should, after a certain time, sell a proportion on the local market so as to enable the Central Bank to meet the demand for foreign currency.

This system did not work as planned and faced with a persistent lack of foreign currency, the monetary authorities reverted to the old formula of 30% for the producer and 70% for the Central Bank. The existence of the SMCP, which continued to centralise the export of frozen seafood, enabled this change to made easily.

The following table shows exports by the SMCP, in value and quantity, from 1986 to 2002 and gives an idea of its importance for the repatriation of foreign currency to the country.

Year	Value of exports (US dollars)	Quantity exported (tonnes)	Average price (US dollars/tonne)
1986	153,356	60,002	2,556
1987	159,833	70,565	2,265
1988	190,779	62,956	3,030
1989	169,399	54,877	3,086
1990	128,400	46,344	2,770
1991	144,841	47,689	3,037
1992	163,455	55,071	2,968
1993	137,694	62,802	2,193
1994	153,684	49,393	3,111
1995	172,534	42,887	4,023
1996	159,992	42,104	3,800
1997	131,300	34,487	3,806
1998	82,797	27,428	3,019
1999	87,643	33,245	2,636
2000	90,111	39,273	2,294
2001	115,847	48,092	2,409
2002	118,188	40,255	2,936

The role of the SMCP in the development of the artisanal fishing

The artisanal fishery has developed greatly on the past few years. The development of the number of pirogues is a good indicator. In 1999 there were an estimated 607 vessels (CNROP/FAO frame survey). Today the number is estimated to be around 3,200 pirogues of different types.

Some artisanal fisheries are seasonal and a vessel or fisher can be involved in different fisheries according to season, opportunities and his know-how.

The following characteristics describe the development of artisanal fishing in Mauritania compared to other segments and explain the interest that Government has shown in it in the various development policies aimed at the fishery sector:
- The catch of the artisanal fishery in 2000 was between 1 and 2% of total catch in the Mauritanian EEZ in quantity terms but was around 5% of the total value.
- Prices received by the artisanal segment (for the same fish) are generally higher than those received by the more industrialised segments due to the better quality of artisanal catch;
- The artisanal segment plays an important role in local landing of catch. Of the other segments, only the national industrial fleet (representing about 12% of total catch by value) lands in Mauritania. The other fleets (EU and others), which represent some 83% of total value, land elsewhere.
- The segment creates many more jobs than the other segments. About 21,380 jobs are created directly or indirectly by the artisanal segment, which is over 72% of total jobs generated by the whole fishery sector;

Although many factors have contributed, the SMCP has played a very important role in the development of the artisanal segment.
- In the early days, the company encouraged new fishers by purchasing their catch even if it was of poor quality, poorly prepared and may otherwise have been unsaleable. It then sold the produce as best it could but often at a loss. In this way the company helped fishers generate the financial resources necessary to improve their equipment and the quality of their produce and hence develop their activities.
- The SMCP also financed artisanal fishing trips so as to reduce their dependence on intermediaries.
- The company provided the means to freeze and store their produce prior to sale by setting up a contract with a local freezing company (SALIMAUREM) at a time when few such facilities existed in Nouadhibou;
- The company also undertook a training campaign for artisanal fishers to enable them to perfect their fishing techniques, improve the quality of their catch, and adapt to an increasingly-demanding export market
- The SMCP arranged contacts between artisanal fishers and foreign buyers, enabling a number of partnership arrangements to emerge

An indicator of the success of this policy is the increasing importance of onshore frozen produce (coming mainly from wetfish trawlers and the artisanal segment) in the SMCP export structure as shown by the following table.

Year	Onboard frozen	%	Onshore frozen	%	Total
1987	64,496	91	6,069	9	70565
1988	51,102	81	11,854	19	62956
1989	43,496	79	11,381	21	54877
1990	37,604	81	8,740	19	46344
1991	37,791	79	9,898	21	47689
1992	40,152	73	14,919	27	55071
1993	40,942	65	21,860	35	62802
1994	30,148	61	19,245	39	49393
1995	26,562	62	16,325	38	42887
1996	24,439	58	17,665	42	42104
1997	20,063	58	14,424	42	34487
1998	11,834	43	15,594	57	27428
1999	15,966	48	17,279	52	33245
2000	17,817	45	21,456	55	39273
2001	23,943	50	24,149	50	48092

Source: Annual statistical bulletin of the SMCP: N° 24 - 2001

In value terms, onshore freezing has increased from 8% of the value of SMCP exports in 1987 to some 52% in 2001 (from around US$13 million to almost US$60 million). And the artisanal product is increasingly sought for its quality.

The role of the SMCP in price formation for Mauritanian produce

The decision to centralise marketing into a single public-service entity like SMCP had as its objective to strengthen Mauritania's market position and to create a capacity to negotiate with buyers who were often organised as pools. It was hoped that the export price would thereby be increased. In fact, operators in the sector were already aware of the need for such an organisation prior to the creation of the SMCP.

The following table shows the development of SMCP exports over the period:

Year	Average price (US dollars/tonne)
1986	2,556
1987	2,265
1988	3,030
1989	3,086
1990	2,770
1991	3,037
1992	2,968
1993	2,193
1994	3,111
1995	4,023
1996	3,800
1997	3,806
1998	3,019
1999	2,636
2000	2,294
2001	2,409
2002	2,936

Without extensive modelling it is difficult to work out the precise impact of the SMCP on these prices. However, a number of observations can be made.

If the SMCP had not been created, or if the State were to remove its monopoly, the most financially-powerful producers would almost certainly create one or more similar structures. Such initiatives would be beyond the reach of all but the most powerful, small and medium-sized producers would be reduced to a satellite role around the new structures.

Whatever the limits to the system discussed above, an important role played by the SMCP has been to maintain a minimum export price at any given time, preventing problems that might have arisen from the dual nature of the system.

It should also be recalled that Mauritanian exports go to a limited number of customers (Japanese and European) and that buyers often operate as a pool. If supply were dispersed, it may become an easy target for powerful financial groups.

The role of the SMCP in developing a fishing sector database

The data concerning fish exports can from two main sources:
- Customs: they supply data which present exports by product type and by value. Products are classified by broad fishery (cephalopods, demersals, crustaceans, pelagics, salted and dried, fish meal, fish oil).
- SMCP provides data on cephalopod exports by market destination and in different categories. These data are published in an annual bulletin but are also available (for instance, for research purposes) at a disaggregated level.

By centralising the marketing of demersal species which represent a large percentage of total exports by value, the SMCP improves knowledge of national fish production and hence contributes to research and resource conservation activities.

SMCP data is currently the best available on Mauritanian production and export of demersal species. It is used as the basis for policy development.

Conclusion

Over the past 20 years, the SMCP has undoubtedly played a positive role in the process of managing Mauritanian fisheries. It has contributed significantly to the development of the exporting segment by taking responsibility for marketing fish abroad.

The company has enabled a minimum price to be held and has prevented supply from being atomised in the face of organised demand. It has collected taxes and duties for the State and in this way has made a significant contribution to funding the State budget. Whatever the fiscal system in place, the SMCP has remained an indispensable tool for tax and duty collection. The SMCP has also played a key role in repatriating foreign exchange.

The company has contributed significantly to the development of the artisanal sector initially by buying the artisanal produce regardless of its quality, then by training fishers and helping them to improve their equipment through interest-free credit.

Finally the SMCP database is the best source of information on fish exports.

Co-Management and Community-Based Fisheries Management Initiatives in Shetland

John Goodlad[1]

Most fisheries around the world appear to be poorly managed. Declining catches and reduced profitability characterise many fisheries, both in the developed and developing world. Recent history suggests that, if fisheries are particularly badly managed then stock collapse can occur leading to economic, social and environmental problems. Two recent examples are the North Sea herring fishery, which was closed from 1975 to 1982 and the Newfoundland cod fishery, which was closed in 1992. Both closures were a direct consequence of poor fisheries management that resulted in stock collapse. While the North Sea herring stock eventually recovered following the closure, the Newfoundland cod stock appears to be showing little signs of recovery despite the 10 year fishing ban.

It is relatively easy to identify those fisheries that are not managed successfully. They are characterised by overcapacity, overfishing and reduced profitability. How can successful fisheries management be defined? A balance between fishing effort and fish stocks, stable stock levels and a level of profitability allowing the industry to invest and operate without subsidy are generally regarded as principal indicators of a well-managed fishery. Another indicator should be how well fishermen and fishing communities are involved in establishing and operating the fisheries management system.

One of the main criticisms of traditional fisheries management is that fishermen and fishing communities feel both neglected and ignored. Decisions on how to manage the fishery are often taken on their behalf by distant Governments and fishing communities do not feel that they have any stake in the fisheries management process. Small wonder then, that under these circumstances, observance of fisheries regulations is less than perfect. Fishermen often feel the fisheries management system they have to operate within has been imposed on them with little or no consultation. There is no sense of ownership and there is no real incentive for fishermen to fully or positively participate in the management of these fisheries.

---

1   Senior Advisor, Shetland Fish Producers Organisation Ltd., and Chief Executive, Shetland Fishermens Association.

In response to these criticisms, attempts have been made in some parts of the world to involve fishermen and fishing communities in the fisheries management process. The principle of subsidiarity, whereby decisions should be taken at the lowest practical level, can be applied to managing fisheries. In some fisheries, particularly those where local stocks are fished, virtually all the fisheries management decisions could be taken locally. In other fisheries, where stocks are distributed more widely, shared or co-management systems between local and central authorities could be introduced.

There are many examples of local fisheries management, co-management and devolution of fisheries management throughout the world. One of the most exciting developments is the creation of Regional Advisory Councils (consisting of fishermen and other stakeholders) which are being introduced as part of the reform of the EU Common Fisheries Policy (CFP). The one area of the EU that probably has most experience of local fisheries management is the Shetland Islands. This paper provides a case study of three fisheries that are currently being managed locally by the Shetland community. Despite the many and varied problems of the CFP, within which all these three fisheries must operate, it seems that these local and community based fisheries management initiatives within Shetland have largely been a success.

Shetland

The Shetland Islands are often described as remote, barren and peripheral. From a fisheries perspective, however, Shetland can be regarded as dynamic and innovative fisheries community of significant regional importance. In fisheries terms Shetland is anything but remote or peripheral. Shetland is actually central to many exciting new developments in European fisheries, not least of which are the local innovations in fisheries management.

Fish and fish products account for over 80% of all Shetland exports while some 20% of the work force are employed in the seafood industry. Around 150 commercial vessels catch around 90,000 tonnes of pelagic, demersal and shellish species. The annual value of this catch is around £45 million. Fisheries dominate the Shetland economy and this at least partly explains the real concern by the wider Shetland community for the well being of the fishing industry within the islands. As well as providing the engine for the local economy at present, fisheries are recognised as being one of the few industries that can flourish in the future in such a remote island community. Shetland has few resources apart from the rich fishing grounds that lie off its shores and it is the desire to conserve these resources for the future that has largely driven these community based fisheries initiatives.

The first innovation is the management of a small industrial fishery for sand eels through a somewhat unlikely local partnership of commercial fishermen and environmentalists. The second innovation in fisheries management is the Shetland Regulating Order, which has been established to locally manage the shellfish fishery. The third innovation is the community ownership of fish quotas in the large demersal fishery that is the mainstay of the local fishing industry.

The fishermen of Shetland, through their two representative organisations (the Shetland Fishermen's Association (SFA) and the Shetland Fish Producers Organisation Ltd (SFPO)) were responsible for initiating these three innovations and continue to play a central role in managing all three fisheries. Each local fishery management system will be reviewed in turn.

The Sand Eel Fishery – Consensus Instead of Conflict

Industrial fisheries have always been contentious. Throughout the world the commercial fishing industry and the environmental lobby generally hold conflicting views regarding the future of fisheries where the catch is used to produce fish meal and oil as opposed to human consumption products. In 1974, Shetland fishermen began to catch sand eels for conversion to fishmeal and oil. By 1982 this fishery had grown dramatically with more than 20 local boats catching 52,000 tonnes of sand eels.

Conflict
During the late 1980's the fishery began to decline. This decline was associated with increased chick mortality in a number of Shetland seabird colonies. A fierce debate ensued between a number of environmental organisations that argued that chick mortality was a direct consequence of the industrial fishery, and the fishing industry, which argued that the chick mortalities and the decline in fishery were both related to a decline in stock. In 1991 the fishery was closed in response to scientific advice that indicated that the Shetland sandeel stock (which is recognised as a separate sub-stock of the North Sea sand eel stock) had reached a very low level. By 1995, the sandeel stock appeared to be recovering and the Scottish Office recommended a limited reopening of the fishery. The fishery was reopened in this year on the basis of a 3,000 tonne per annum quota with a closure at the end of June in order to ensure that no fishing took place when seabird chicks were being fed. The Scottish Office management measures were a compromise between the conflicting viewpoints. Shetland fishermen had wanted to see a higher quota and no closure, while environmental bodies had argued for the fishery to remain closed altogether and, at worst, a full closure during the whole of June and July. The Scottish Office found that its management regime satisfied neither side and was heavily criticised by both the environmental groups and the seafood industry.

Consensus
And then the unthinkable happened. Two of the environmental organisations, which had been very critical of the fishery, the Royal Society for the Protection of Birds (RSPB) and Scottish Natural Heritage (SNH), met with local fishermen in 1997 to discuss the possibility of an improved management scheme for the fishery. After

extensive negotiations, it was agreed that there should be an annual fishery with an initial quota set at 7,000 tonnes (which would be reviewed annually). The fishery would be completely closed during the whole of June and July. Moreover, the fishery would be managed by the local fishing industry, as opposed to the Scottish Office. It was also decided that no vessel over 20 metres would be allowed to participate.

The 1998 Shetland sandeel fishery was successfully managed on this basis and the fishery has been locally regulated ever since. While the management framework was set by central Government, the effective day-to-day regulation of the fishery has been undertaken by the fishermen of Shetland (through the SFPO) in consultation with RSPB and SNH. From 1998 to 2000, the numbers of vessels allowed to participate in the fishery was limited to those that had a historical fishing record of fishing for sand eels and individual per vessel quotas were set to ensure that the 7,000 tonne quota was not exceeded. These regulations were not set out in the management framework (unlike the 20-meter vessel length limit and the close season) but were introduced by the SFPO as local regulations necessary to properly manage the fishery. While enforcement of the management framework is the responsibility of the UK Government (through the Scottish Office), the enforcement of the additional local regulations was the responsibility of the SFPO. This was not a problem as there has always been widespread support amongst fishermen that local management of this fishery is much better than regulation by the Scottish Office.

These local regulations have been relaxed in recent years, as it became clear that the annual quota would not be taken. The annual catch has now fallen to around 2,000 tonnes and only a few boats participate. It is generally recognised that the sand eel stock has declined but not as a result of overfishing. Since the fishery has been reopened it has been strictly managed in accordance with scientific advice. Indeed, the actual catch has generally been significantly less than the annual quota. It appears that the sand eel stock (which is naturally very variable in size given its dependence on only a few year classes) has declined because of environmental or other factors.

Despite the fact that this fishery is now only prosecuted by a few vessels (with the result that little or no local management is now actually required), the local management system remains in place should this fishery again increase. The management framework is the result of an agreement between environmental groups and commercial fishermen and is not something that has been imposed by central Government. It therefore has the support of all stakeholders in this fishery. The fact that the day-to-day management of this fishery is the responsibility of the SFPO, in consultation with RSPB and SNH, further enhances the sense of local ownership of this local fishery.

The Shetland Regulating Order

The Shetland shellfish industry is worth around £3 million per year to the 100 or so inshore vessels, which participate in the fishery. The principle species caught are lobster, crab (brown & velvet), scallops (king & queen) and buckies. Apart from a general UK fishing vessel-licensing scheme, there has never been any specific regulatory framework for the management of the shellfish industry on a local or regional basis. The lack of effective regulation has resulted in over exploitation, stock decline and a consequent reduction in earnings. The local nature of shell fishing strongly suggests a regional approach to regulation. The SFA tried unsuccessfully for many years to persuade the Government to introduce and administer a local shellfish-licensing scheme for Shetland waters. It eventually became clear to the SFA that the Government had neither the inclination nor desire to introduce such a regional licensing system for the shellfish industry. Falling catches and a growing alarm at falling earnings prompted the SFA to take the initiative and try to establish a local management system.

The SFA looked at the possibility of using existing legislation – in this case, the Sea Fisheries Act of 1965, which provides for the establishment of Regulating Orders. A Regulating Order enables a fishery to be locally managed by an organisation set up for the purpose of conserving the stock and improving the fishery. To date only a few Regulating Orders have been introduced, usually for specific fisheries in particular areas (e.g. river estuaries). The Shetland proposal was to introduce a comprehensive fisheries management scheme for the entire Shetland shellfish industry. The initial response from Government was that this was far too ambitious and that Regulating Orders were really intended to manage single species fisheries in very well defined local areas. The SFA were not easily deterred and embarked on a series of meetings throughout Shetland to ensure there was maximum support throughout the wider community for the principle of local management. Having confirmed that the support for local management was indeed very strong and widespread throughout the islands, the SFA resolved to press ahead with the Regulating Order proposal.

The SSMO

The first step was to establish a local management group. Although the SFA took the initiative, it was realised that the management process had to involve a much wider group of stakeholders. As a result the Shetland Shellfish Management Organisation (SSMO) was set up in 1996. The SSMO consists of representatives of the local Government (2), local Community Councils (1), shellfish processor (1), an environmental organisation – Scottish Natural Heritage (1) and the local Fisheries College (1) as well as the SFA (4). Although the SFA is the leading organisation with 4 seats on the SSMO, it does not have a majority of votes. Neither does the SFA hold the Chairmanship of the Organisation, the post being held by a Fisheries biologist from the local Fisheries College. In this way the SSMO can rightly be regarded as an organisation that genuinely represents the wider community and includes all stakeholders in the fishery. The SFA took the decision that, if an application to establish a Shetland Regulating Order was to be successful, the approach had to be made on behalf of the wider Shetland community.

The SSMO set out its vision of managing all shellfish stocks within 6 miles of the Shetland coast in its application to establish a Shetland Regulating Order. The clear objective was to conserve all shellfish stocks so that the sustainable exploitation of these stocks could be ensured in the future. Unless all species were included, the SSMO argued, fishing effort would simply be redirected onto those species outwith the scope of the Order. In order to regulate all shellfish stocks, the Sea Fisheries Act had to be amended as the original Act only referred to mussels, oysters and lobsters. The Act had to be changed to include the various crab, scallop and whelk species. Although a quite simple piece of legislation, it was difficult to find parliamentary time at Westminster – the management of the Shetland shellfish industry was simply not important enough to secure valuable UK parliamentary time. A solution to this problem was found when the Scottish Parliament was created in 2000. Regulation of inshore fisheries was a devolved matter and the newly created Scottish Parliament was able to find the time to amend the original Act, insofar as it applied to Scotland, by including crabs, scallops and whelks in the list of species for which a Regulating Order could be introduced.

The 6-mile limit is the maximum geographical area within which a Regulating Order can be introduced. As far as Shetland is concerned this was not a problem since most shellfish stocks are located within the 6 mile limit.

The SSMO proposal was based on the fundamental principle of effort limitation. Initially all fishermen with a historic performance in the Shetland shellfish fishery (including some non-Shetland fishermen) would receive a Shetland permit. Following this initial allocation, new permits would only be issued if shellfish stocks could sustain additional fishing effort. In reality it was recognised that fishing effort was likely to reduce (at least in the short term) as fishermen retired or withdrew from the fishery and these permits were not re-issued. In order to achieve this reduction of effort, it was recognised that trading of Shetland permits had to be prohibited. Whenever a vessel no longer needed a permit, it was resolved that the permit had to be returned to the SSMO.

In addition to the permit system, the Regulating Order proposal also provided for additional management and technical measures covering vessel size, gear type, closed areas and seasonal closures.

Finally, it was proposed to initiate stock enhancement schemes such as lobster restocking. However, it was pointed out that such restocking schemes could only be effective if a proper regulating framework was in place. In other words the introduction of the Shetland Regulating Order was clearly the first step.

After being set up in 1996, SSMO submitted its proposals to Government in January 1998. Following almost two years of discussion and debate, the Scottish Executive eventually established the Shetland Regulating Order in March 2000. Over 170 permits were initially issued to all commercial vessels as well as a large number of part time operators. In addition to the central management measure of effort control, the SSMO has also introduced a variety of additional management measures. A maximum vessel length for crabber vessels has been brought in, a maximum limit on the number of scallop dredges each vessel can use has been stipulated as has a prohibition on dredging for scallops during the hours of darkness, a close season has been established for velvet crabs and a minimum landing size introduced for certain species.

Given the enormous effort that the Shetland community at large put into the campaign, the decision to establish the Shetland Regulating Order has rightly been regarded as a major success within the islands. It has been created from the "bottom up" not imposed from the "top down". It is effectively local fisheries management with minimal interference from the EU Commission or the UK Government. The management of the Shetland shellfisheries is now undertaken by the people who participate in the fishery and by the wider community of stakeholders. While it is too early to claim that the Shetland shell fish stocks have been restored to former levels of abundance, it can be demonstrated that, for the first time ever, total fishing effort is now controlled and has actually been somewhat reduced over the past two years. Moreover, the additional technical measures (such as maximum vessel length and minimum landing sizes) are also beginning to have a positive impact on shellfish stocks. Most importantly of all, this is being achieved through consensus and general support throughout the fishing community.

Funding and Enforcement
While the Shetland Regulating Order has been a major success in terms of better fisheries management, there have been some problems. The issues of funding and enforcement have been particularly difficult. With regard to funding, with responsibility for fisheries management being devolved to the SSMO, the full financial cost of management has also been devolved. The price of local management has been financial responsibility. The SFA, together with some of the local stakeholders, initially provided funding to establish the SSMO office, set up the administrative system and employ a full time staff member. The intention was that the ongoing operating costs would be funded through a permit fee. In reality the income from permits is insufficient to cover all the ongoing administration costs. The permit fee that would be required for full cost recovery would be too high, especially for the smaller boats, many of which are owned by part time fishermen. This financial problem was not however unexpected. After all there are very few fisheries management systems in the world where the commercial fishery funds the entire cost of fisheries management. In most fisheries there is a large level of public sector funding. There is no reason why Shetland will be any different and discussions with local public sector funding agencies in Shetland are currently ongoing to secure the modest level of public sector funding required to cover the operating deficit.

The problem of enforcement is rather more serious. In the same way as full financial responsibility has been devolved, so the enforcement of the regulations has also become the responsibility of the SSMO. The resources of the Scottish Fisheries Protection Agency (SFPA) have not been made available to the SSMO. The SPFA staff and fishery protection vessels cannot be used to monitor individual vessel compliance with the regulations. Neither can the SFPA resources be used to prepare a case for prosecution to the Courts. The SSMO must enforce its regulations entirely from its own resources. Given its ongoing funding problems, the capital and revenue costs of fishery patrol vessel and fishery officers are far beyond the resources of the SSMO. In the same way the legal costs of preparing a private prosecution would be considerable and would again be beyond what the SSMO could afford. Once more it can be argued that it is somewhat unreasonable to expect a small commercial fishery to totally fund all these enforcement costs without any public sector support whatsoever.

The enforcement problem could have resulted in the collapse of the Shetland Regulating Order. With inadequate resources to effectively enforce the regulations, a number of breaches of the rules without an appropriate enforcement response would have led to widespread abuse of the system and the collapse of the authority of the SMMO. This has not happened because of the fact that the fisheries management system has been drawn up by, and is administered by, the fishermen and other stakeholders. The system continues to command widespread support within the Shetland and peer group pressure has been extremely effective in ensuring continued compliance with the regulations. This more than anything else illustrates why local fisheries management works better than the management of fisheries by distant and remote administrations. Even without public sector funding to cover enforcement costs, the Regulating Order has achieved a remarkably high degree of compliance from the permit holders. Put simply, this is because the permit holders (through the SFA) have been closely involved in drawing up the regulations and there is a real sense of ownership by individual fishermen. It is however clear that an effective enforcement system will have to be established before too long and a number of options are currently being considered.

The Shetland Regulating Order is the only local management system of its kind within the UK. If the early success of the Order is continued over the next few years, it is likely that other areas of the UK will use the Shetland Order as an example of how to best manage local shellfish stocks. There is no reason why other communities cannot manage their shellfish fisheries as successfully as the Shetlanders can. The key to success is that the management system is local as opposed to national.

Community Fish Quota Scheme

Before 1984, demersal and pelagic fisheries within the UK were managed on the basis of fortnightly or monthly quotas that were allocated to individual vessels by Government Fisheries Departments. For example, if North Sea whiting or West of Scotland herring were subject to catch limits, all UK fishing vessels would receive the same fortnightly or monthly quota from Fisheries Departments. From time to time these quota allocations varied depending on vessel size. Such a system took no account of regional variations or of the requirements of different sectors of the UK fleet. This system was also rather remote in that fishermen, through their organisations, were not directly involved in the decision making process as such. There were regular consultations with the fishing industry, but the final decisions on setting vessel quota limits were made by Government. It was a classic case of distant decision-making where fishermen no real sense of involvement.

The absence of any real involvement in the decision making process, together with the lack of a regional dimension in the quota allocation process, led to much criticism. Nowhere was this criticism more marked than in Shetland. As already outlined, there was a lucrative industrial fishery for sand eels around Shetland during the 1980's – especially during the summer months. This fishery regularly attracted a large number of vessels that would otherwise have been catching demersal fish. This left only a small number of trawlers to supply the local fish processing plants. In 1983 the UK haddock quota limits were particularly poor during the summer months. One unforeseen result of this quota level was that the reduced numbers of Shetland trawlers were unable to land enough haddock to supply the needs of the local fish processing industry. By the time the sand eel fishery had finished in September, the haddock catch limits had been raised. But, although the entire Shetland trawler fleet was now able to fish for haddock, the 'summer haddock fishery' had been lost. The possibility of landing sufficient haddock during the summer months, when a large proportion of the fleet usually diverted to industrial fishing, only seemed possible if larger per vessel quotas could be allocated, something which was patently impossible under the national quota system which existed at this time.

In view of this, the Shetland fish catching and fish processing industries argued that a more flexible system of quota management was necessary in order to take account of the particular circumstances pertaining in Shetland at this time. The SFPO had been established in December 1982 in order to try and improve the marketing of its members' catches. Since the effective marketing of the haddock catch to the local fish processing industry was being prevented by an inflexible quota management system, the SFPO began to promote an alternative system.

Why not, it was suggested, allocate to the SFPO that share of the UK haddock quota that the Shetland fleet would normally catch during a full year? This quota could then be shared between member vessels in the manner best suited to local market conditions. The Scottish Office Fisheries Department was persuaded and the SFPO received its own haddock quota allocation for 1984. The first tentative steps had been taken to decentralise quota management within the UK. This system of allocating quota to fishermen's organisations became known as the sectoral quota system.

This experiment, in so far as the Shetland fishing industry was concerned, was successful. Unlike the previous system, this system allowed fishermen, through their representative organisations, to manage fisheries. At the same time Fisheries Departments welcomed the opportunity to devolve the increasingly burdensome and time-consuming job of fisheries management. As a result there was a rapid move towards adopting sectoral quotas throughout the fishing industry at large. By 1985 most haddock, cod, whiting and saithe fisheries were being managed under this system. Within two years the herring and mackerel fisheries were also being managed in this way. While the UK Government continued to retain a significant role in fisheries management, much of the day-to-day management had been devolved to fishermen's organisations. This is an early example of successful co-management.

Producer Organisations
Producer Organisations (POs) are a relatively new type of fishermen's organisation. Unlike the long-established trade organisations (many of which, like the SFA have been in existence for over 50 years), POs are the direct result of British membership of the EU. The first British PO was established in 1973 – the year Britain joined the then Common Market.

Under the terms of the EU Common Fisheries Policy (CFP), POs play a central role in the common organisation of the market. The principal objectives of the POs throughout Europe are 'to encourage rational fishing and to improve conditions for sale of their members' products'. In order to achieve these objectives all European POs have a responsibility to implement the marketing regulations of the CFP. The PO system enables fishermen to enjoy the benefits of the EU minimum price scheme and market support mechanisms. POs therefore must ensure that fish landed by member vessels are properly graded according to EU size and freshness criteria. The EU official withdrawal price (i.e. the minimum price below which fish cannot be sold) must be strictly observed if PO member vessels are to benefit from the market intervention system (i.e. the system whereby financial compensation is paid for fish which cannot be sold at the official withdrawal price).

All POs throughout Europe are now involved, to a greater or lesser extent, in the implementation and administration of the EU marketing regulations. Some POs have become involved in related activities such as the establishment of quality control systems, the marketing of fish and the establishment of fish processing plants. It is only within the UK, however, that POs have come to play a central role in fisheries management. This new role for POs was recognised in 1993 when the EU marketing regulation was amended to allow POs, at the discretion of member states, to manage national catch quotas. With this change in the relevant regulation, the EU has clearly signalled its approval of fisheries management by the PO sector. It will therefore be interesting to see if other member states follow the UK approach and develop fisheries management systems based on PO participation.

Within the UK there are now a total of 19 POs. These are largely, although not entirely, regionally based. These POs now represent the vast majority of Britain's fishermen, boats and catch. In terms of fish quotas it is estimated that the PO sector now manages over 95 per cent of all quotas. The 19 POs reflect the geographical and sectoral diversity of the British fleet.

That proportion of the British fleet that is not in membership of the PO sector is referred to as the 'non-sector'. Although the non-sector accounts for a fairly large number of boats, in quota terms it represents less than 5 per cent of British quotas. The non-sector is managed in much the same way as all fisheries were managed before 1984, i.e. by individual vessel monthly allocations set by Fisheries Departments. The non-sector largely consists of smaller vessels, but in addition to the non-sector, there is also a very large number of 'under 10 metre vessels' within the UK fleet. While very large in terms of numbers and individual vessels, this sector obviously consists entirely of small boats, many of which are operated on a part-time or seasonal basis, and is also managed directly by Fisheries Departments.

Since its introduction in 1984, the sectoral quota system has continued to develop and adapt to changing circumstances. Each year Fisheries Departments issue a consultation paper, which proposes certain administrative or organisational changes to the basic system. On the basis of the response from the POs and the non-sector, changes are made to the management system for the ensuing year.

Through regular changes and modifications, the sectoral quotas have now developed into a complex and comprehensive system of fisheries management. POs must now manage all fisheries for which there are UK quotas in the North Sea (ICES Area IV), West of Scotland (ICES Area VI) and the Irish Sea and English Channel (ICES Area VII). It had previously been possible for POs to manage some fisheries and yet to opt to remain under non-sector management (i.e. management by Fisheries Departments) for others. Sectoral quotas have been calculated on the basis of the actual catches (track record fishing performance) of member vessels during the previous three years. Pelagic quotas had previously been based on the catches of the previous two years. Since 1992, track record fishing performances have been attached to, and transferable with, vessel licences, rather than to the vessels themselves.

Having established the sectoral quota available to each PO, Fisheries Departments then monitor catch uptake and will close a fishery when the sectoral quota has been caught. Apart from this, the management of the quota is largely at the discretion of the PO concerned. As one would expect, different POs adopt different quota management strategies depending on their individual circumstances. In so far as demersal fish quotas are concerned, most POs continue to allocate quotas to member vessels on a monthly basis. Sometimes these quotas vary according to vessel size bands, but more often similar quotas are allocated to members regardless of vessel size. Several POs, however, began to allocate individual annual vessel quotas based on the track record fishing performance of individual vessel licences. In other words the quota allocation that a PO receives on behalf of a member vessel is simply reallocated to that vessel on an annual basis. Some POs allocate all their quotas on this basis, others only for some species and others only for some vessels. Other POs have made similar arrangements but on an individual company basis as opposed to an individual vessel basis.

Pelagic quotas are essentially allocated to only two POs (the main Scottish PO and the Shetland PO) with individual pelagic quotas being allocated to those pelagic vessels not in membership of these two POs. Both POs in turn allocate individual annual pelagic quotas to their pelagic members. Again these quotas are related to individual track record fishing performances.

A further development of the sectoral quota system was to allow POs complete discretion in swapping fish between each other. This ensures that UK quotas should not remain uncaught. Quota swaps are now becoming very commonplace with POs swapping away fish quotas unlikely to be caught in return for fish quotas that are in short supply. Direct fish for fish quota swaps are most usual although quota gifts (which can be re-paid in future years) have become more common.

In other words there is a huge diversity of quota management systems, both between PO's and within PO's. The fact that this diversity has developed illustrates more than anything else why a single, uniform, centralised management system can rarely meet the needs of a diverse, varied and regionally based fishing industry. Co-management in contrast can deliver a wide range of fisheries management solutions, each reflecting the local, regional and sectoral issues pertaining to each fishery.

Purchase of Fish Quotas

Purchase of fish quotas has in effect been possible since 1993 through a special scheme that was introduced to complement the fleet decommissioning scheme introduced at this time. Under this scheme, vessel owners interested in decommissioning their ships could, as an alternative, sell their quota entitlement (arising from their track record fishing performance) to the PO they had been in membership of for the past three years. In return for selling quota entitlement to the PO, the vessel owner had then to relinquish his vessel licence in the same way as when a vessel owner accepts a decommissioning grant. This scheme is essentially a form of industry funded decommissioning, and since the requirement to have been in membership of a PO for three years was subsequently relaxed, a vessel's quota entitlement can now be sold to any PO. The SFPO has purchased the quota entitlements of seven separate vessels to date, with the express intention of retaining these quota entitlements by the organisation as a whole.

As already noted, Fisheries Departments continue to manage the non-sector. Over recent years the size of the non-sector has been reduced. Most larger vessels are now in membership of POs and the non-sector now consists of a large number of small vessels. The departure of larger vessels from the non-sector to the PO sector has resulted in substantial track record fishing performance being lost from the non-sector. This has in turn resulted in the non-sector quota allocations being further reduced with very poor per vessel quotas being allocated to non-sector vessels by Fisheries Departments.

As the sectoral quota system developed so did an increasing awareness of the importance of track record fishing performance. As each PO endeavoured to maximise its sectoral quota allocation, increased attention was focused on the catch record of vessels applying for membership. Most POs adopted a policy of only admitting into membership those vessels that had a track record fishing performance comparable to vessels of a similar size already in membership. It became widely recognised that admitting vessels into membership with inadequate catch records would simply result in new members requiring quota allocations but unable to contribute significantly to the quota pool.

The non-sector consequently became a residual for that proportion of the fleet that had very poor track record catches and were therefore unable to secure membership of a PO. Real fears were expressed that, if more of the non-sector fleet with reasonable track record catches were to join the PO sector, the already poor non-sector quota allocations would become even worse. In response to this, Fisheries Departments have, since 1995, guaranteed a minimum quota allocation to the non-sector.

Throughout the development of sectoral quota system, UK fishermen remained almost unanimously opposed to the buying and selling of quotas. The SFPO quota purchase initiative was therefore roundly criticised by many UK fishermen. In response, the SFPO pointed out that it was simply trying to secure future fishing opportunities for member vessels and that this option was open to all UK PO's under the system rules. While no other PO followed the Shetland example, increasing numbers of fishermen began to consider securing additional quota on an individual basis.

One of the main advantages of the management system, based on the three-year rolling reference period, was the fact that quota allocations were directly based on the historical landings of vessels. In this way quota allocations would always bear a close resemblance to actual landing patterns. The principle disadvantage was however the scope for individual vessels to increase their track record fishing performance by deliberately increasing their landing records. In other words fish that had not been caught were "landed and sold" in order to improve a vessels track record during the reference period. This became known as "ghost fishing" and is, of course, the opposite of over quota or "black fish" landings. The scale of "ghost fishing" increased during the mid 1990's as more and more fishermen realised that fishing vessel licences with large track records were worth more than licences with small track records.

In an effort to prevent "ghost fishing" spiralling out of control, UK Fisheries Departments and the fishing industry established a Working Group to examine this issue. The Groups concluded its work in 1997 and recommended that, in future, the management system should be based on fixed quota allocations derived from vessel landings made during the three-year reference period of 1994, 1995 and 1996.

This would, the Group argued, fix track records and thereby remove the incentive to "ghost fish". It would also simplify the quota allocation system in that every vessel licence would have a fixed track record attached that would not be subject to annual variation depending on catch performance.

Fixing track records, detractors argued would remove the flexibility inherent in the rolling reference period. It would also, it was argued, take the industry closer to a system of property rights based on Individual Transferable Quotas (ITQs) something that had hitherto been opposed by British fishermen.

In the event, UK Fisheries Ministers accepted the Working Group recommendations and have managed UK fisheries in 1999 on the basis of Fixed Quota Allocations (FQA's). Each vessel licence has been allocated a FQA based on its landings from 1994 to 1996. This allocation is expressed by species in 100kg units and, when related to the total UK quota during this period, is effectively the individual vessel quota entitlement of the UK quotas. Although the FQA is a fixed unit, it is worth more than 100 kg. when quotas are high and worth less when quotas are low. With the decline in North Sea cod and haddock stocks in recent years, the "fish value" of an FQA of these species is worth has declined considerably.

Apart from the change from a rolling to a fixed reference period, the sectoral quota system has remained unchanged with quota allocations being made to POs on the basis of member vessels FQA's. In reality, however, a fundamental change had been made to the management system that focused the attention of licence holders on the size of individual quota allocations and their value in terms of selling and renting. The UK fishing industry had consequently taken a huge step towards a management system based on property rights.

The Legal Position

This debate on property rights within the UK fishing industry has given rise to an interesting legal question. It is clear that independent nation states (such as Iceland and New Zealand) can confer a legal property right of fish quota to individuals out of what has hitherto been regarded as the property of the nation state. Within the EU, the situation is different. Member states do not "own" fish quotas as such. Fish quotas are essentially a "common resource" under the terms of the CFP. Each member state receives an annual entitlement to fish a proportion of this "common resource". An agreed allocation key that is known as relative stability determines these national allocations. While these national allocations are the responsibility of each member state to manage (and hence the variety of fisheries management systems adopted within the European Union) the national allocation is not the "property" of the Member State as such.

There is therefore a very interesting debate as to whether or not the conferring of property rights can be done by the member states or could only be done by the European Union itself. If property rights were to be conferred by the European Union, and these were to become freely tradable amongst community fishermen, the principle of relative stability would be compromised. This would cause enormous political problems for many member states and would therefore be resisted. On the other hand, member states may not have legal competence to confer property rights on their own fishermen. Until this question is resolved, it is unlikely that British, or other E.U fishermen, will be able to obtain full property rights over fish quotas

The Industry Response

Although UK fish quotas are not assets with legal title of ownership (as is the case with property rights), for some time now fish quotas have been seen as having a monetary value. Ever since POs allocated individual quotas to individual member vessels (based on that vessels track record catch), fish quotas have become tradable. Although the change to FQAs did stop the deliberate enhancement of track record performance for subsequent resale, it also focused attention of all fishermen on what their track record fishing performance actually was. For the first time, all UK fishermen had to confirm their agreement with the quota allocations attributed to their licence. This raised the profile of quota value in a way that had never been the case under the old management rules.

The fact that fish quota is not a legal asset in its own right, and is inextricably linked to a vessel licence, has not discouraged the emergence of quota trading as a growing activity. Standard legal agreements are now used to separate quotas from licences and the industry itself has argued that Fisheries Departments should formally reallocate individual FQAs every year to reflect the quota trading which has taken place during the preceding year. This was agreed in 2001 and a regular review of individual FQA's now takes place every few years. This review formally revises individual FQA's to take into account any quota trading which has taken place in the intervening period.

Most quota trading has taken place in the demersal fish sector; the pelagic sector having fewer vessels and a much smaller number of quota transactions. The clear trend has been the purchase of additional quota by the more successful fishing partnerships and companies that have in turn made arrangements with their PO for an individual annual quota to reflect the enhanced catch opportunities which they have purchased. This has inevitably hastened the decline in fishing vessel numbers as less successful partnerships, or licence owners wishing to retire, have sold their FQAs.

In addition to the sale of FQAs, quota rental is now becoming more widespread. Fishermen, who are perhaps short on quota during a quota period, or lack the capital to purchase FQA, often rent quota from those with quota surplus. Again, because of the fact that quotas do not have separate legal title, rentals are subject of standard legal agreements involving the 'swapping' of fish between POs.

The development of a market for quota rental has in turn encouraged the emergence of a group of FQA holders who have decided it is easier to rent quota rather than fish it. This is a development that appears to be commonplace wherever ITQs have been introduced.

The inevitable consequence of the developing market for quota purchase and quota rental is a call for the system to be further simplified whereby the quota would be separated from the licence and would thereby assume a legal entity of its own. This would eliminate the need for costly legal agreements covering quota purchase and quota rentals, but would of course be a change which would further transform the sectoral quota system and move it even closer to a system based on property rights.

Despite the ever increasing scale of quota trading which is now taking place, the official policy position of the UK national fishermen's federations is still one of opposition to ITQ's. This reflects the fact that the majority of fishermen have not yet participated in the quota trade. But a sizeable, and ever growing, minority have bought, sold or leased quota. These fishermen are becoming increasingly vocal in defending the system of quota trading. There are now clear signs of the beginnings of a fundamental shift of opinion amongst British fishermen regarding the issue of property rights.

The Shetland Response

The Shetland PO recognised at an early stage that quotas were becoming valuable commodities. As early as 1993, the SFPO took advantage of the Government scheme to purchase fish quota in association with the decommissioning of vessels/licences. As already noted, the SFPO have purchased (between 1993 and 1997) a total of seven vessel quotas under this scheme. The SFPO took a decision to make a substantial investment in purchasing fish quotas under this scheme with the aim of holding this quota in corporate ownership for the benefit of all member vessels, both at present and in the future. This total investment amounted to almost £1 million enabling the Organisation to purchase 2,386 tonnes of demersal fish. With the introduction of FQA's this quota purchase subsequently became 23,860 FQA units.

As quota trading has become more widespread, the cost of acquiring quota has increased. This has been particularly marked in 1998, as a result of the introduction of FQAs in 1999. As an indication of how quota values have increased, it can be noted that the SFPO paid £250 per tonne for cod in 1995; by 1999 the cost of buying a tonne of cod has increased almost eight fold to £2,000 per tonne. This increase in the value of quotas reflects what has happened in most fisheries around the world where quotas are traded. The dramatic fall in the quotas of cod and haddock in the North Sea in 2002 and 2003 has resulted in something of an economic crisis within the demersal sector. This has in turn led to a reduction in the level of quota trading and a recent fall in quota prices.

This quota investment by the SFPO has enabled the Organisation to provide all member vessels with enhanced fishing opportunities. Largely as a result of this, few individual Shetland fishermen have seen the need to purchase additional quota in their own right. The Shetland PO has also been unique amongst UK PO's in becoming a quota holder in its own right. With the enormous changes taking place in the UK fishing industry in terms of quota ownership, there is concern within fishery dependent communities such as Shetland that, unless there can be greater security over quota, fish quota could be sold outside the islands never to return. Shetland's economy, after all, is dependent on the fishing industry and, without fish quotas, future economic and employment prospects for Shetland could be bleak. While the initiative, which had been taken by the SFPO, was commendable, this FQA of 23,862 units represented only a small proportion of the total SFPO quota. The risk of Shetland loosing its fish quota was real and the SFPO initiative was by itself insufficient to retain fish quota in Shetland for the future. There was a growing consensus within Shetland that the community should try to secure ownership of fish quotas.

After much debate it was agreed that one of the islands development funds

(Shetland Leasing and Property (SLAP)) should buy fish quota in order to establish a pool of community held quotas. The SFPO agreed to act as agents for SLAP and to hold any quota purchased in a separate "community FQA". A total of almost £17 million was invested in acquiring 117,060 FQA units of demersal fish quota between 1998 and 2001.

The SFPO therefore holds two pools of demersal fish quota; one representing its own investment of 23,862 FQA units with another held by the SLAP, on behalf of the wider Shetland community, of 117,060 FQA units. Taken together, these two quota pools amount to some 140,922 FQA units. This is a significant quota pool compared with the 103,921 FQA units owned privately by the individuals and companies within the Shetland fleet. In summary therefore, out of the total of 244,843 FQA units administered by the SFPO some 140,922 or 58%, is held in community ownership (either by SFPO itself or by the SLAP).

The fact that a significant proportion of Shetland's demersal fish quota is effectively held in common ownership is unique within the UK. But what makes the "Shetland ownership" of the marine resource so significant is the use that is made of this quota.

The New Entrants Scheme
With the development of quota trading, the cost of entering the fishing industry has escalated. Not only does a prospective fisherman have to pay for a boat and fishing licence, he must also now find the cost of fish quota. In many cases the cost of quota is as much as the combined cost of a boat and fishing licence. This has the inevitable consequence of making entry into fisheries much more difficult for young fishermen. There is now a clear trend for existing successful fishing partnerships and companies to acquire additional quota and thereby expand their operations. In short, the UK fishing industry is becoming concentrated into fewer hands as quota is traded.

While Shetland has many successful fishing partnerships, there is no guarantee that their FQAs will not, in the fullness of time, be sold outwith the islands. At the same time, there may be many young Shetland fishermen who lack the capital to buy quota and thereby become successful fishermen in their own right. It was this scenario of an island group, surrounded by fish and dependent on the seafood industry, seeing its marine resource being eroded away which prompted first the SFPO and then SLAP to enter the market and purchase quota.

The SFPO quota pool is used to augment the monthly quota allocations of member vessels and will continue to enhance members fishing opportunities in the future. There are no circumstances in which the SFPO quota pool would ever be sold and it will therefore continue to provide additional fishing opportunities for future generations of Shetland fishermen. This is the "return on capital" which the SFPO is obtaining from its quota investment.

In contrast, the SLAP quota pool is being used to help new entrants get started in the industry. The SLAP quota pool is set aside by the SFPO as a quota reserve for fishermen who cannot afford to purchase quota. These fishing partnerships are able to become full members of the SFPO, and are able to fish out of the general quota pool (and thereby obtain the benefit of the additional SFPO quota pool) despite not having any individual FQA. Instead they pay a proportion of their gross earnings to the SFPO in order to "rent" a share of the SLAP quota pool. To date a total of 8 "new entrants" have been able to acquire a boat and licence and start fishing without having to purchase quota. The total numbers of new entrants able to start fishing under this scheme will ultimately be limited by the size of the SLAP quota pool. The intention is, however, to continue to invest in quota in order to enable more new entrants to join the industry. So far only demersal fish quota has been purchased but there is also the possibility of acquiring pelagic quota in the future.

The "return on capital" for SLAP is the rental income (which the SFPO collects and remits to the SLAP), any future appreciation in quota values and, finally, the fact that investment in these quotas is enabling fishing activity to continue to develop within the islands. As well as creating jobs at sea, an additional Shetland fishing vessel will create employment onshore in the fishing processing and ancillary services.

ITQ's by another name?

The system of UK fisheries management is firmly based on quota allocations and is therefore a classic resource based management system. The development of the sectoral quota system, and especially the recent introduction of FQAs, has resulted in fish quotas being bought, sold and leased. Although there is no legal title to UK fish quotas, increasing numbers of fishermen are prepared to invest in an administrative system that actually confers most of the advantages of an ITQ system. For many people the UK system of fisheries management by sectoral quotas based on FQA'S is virtually a system of ITQs by another name.

At the same time there are important differences. The fact that FQA holders do not have a legal title over their investment is clearly an important difference. In reality, it would however be inconceivable that any Government would consider abandoning quotas as the method of managing fisheries. Another important difference is that POs play an important role in the management of UK quotas. Under a classic property rights system it would be difficult to envisage a role for POs as fishery managers.

The debate over whether the UK should formally introduce ITQ's, with legal title for fish quotas, will continue. The UK Fisheries Minister has rightly said that there is presently no consensus within the fishing industry for such a system. Having said this, increasing numbers of fishermen (who have bought FQAs) are now calling for ITQ's to be introduced. Indeed, there are now the beginnings of what could be a seismic shift in attitudes amongst UK fishermen on the issue of property rights. It may only be a matter of time before the majority opinion favours a system of ITQ's. It is therefore probable that further changes will in due course be made to the sectoral quota system that will separate FQAs from vessel licences and thereby confer legal title on fish quotas.

Whether or not there is a further move towards ITQs, the Shetland system of "community owned fish quota" will continue to secure access to the marine resource for a fishery dependent island community. In particular, this pool of community fish quota will continue to be used to help young fishermen start their fishing careers without having to invest in fish quota. There is no reason why the concept of community fish quota cannot be as valid under an ITQ system as it is under the sectoral quota system.

Assessment

The sectoral quota system has been a success insofar as most British fishermen are concerned. Through this fisheries management system, British fishermen have, for the first time, participated in a programme of co-management whereby the PO's effectively manage the fisheries. While the UK Government continues to have overall responsibility for managing the sectoral quota system and ensuring that no PO exceeds its annual quota allocation (the Government effectively manages the PO's) the PO's have complete responsibility regarding the internal allocation of quota to individual vessels (the PO's manage the fishermen). The system has worked well insofar as the fishermen have become full and active stakeholders along with the Government in the co-management of UK demersal and pelagic fish quotas.

The Shetland community quota scheme has also been a success. The SFPO quota purchase scheme has provided member vessels with proportionally better per vessel quotas than other UK fishermen and to some extent has helped protect Shetland fishermen from the worst ravages of cod and haddock quota cutbacks during the past two years. The SLAP scheme has created a pool of community owned fish that will be fished by future generations of Shetland fishermen. The SFPO and SLAP quotas are a remarkable innovation in community ownership of fish quotas with few parallels elsewhere in the world.

The sectoral quota system has however been less than completely successful when assessed against the objective of stock conservation. Although the annual quotas of many fisheries (e.g.: saithe, herring, mackerel and nephrops) have remained stable or have increased, the cod and haddock quotas have been dramatically reduced, especially over the last few years. The annual UK catch of cod had ranged from 60 to 70 thousand tonnes during the 1990's, while UK haddock catches varied from 55 to 90 thousand tonnes during the same period. This catch had fallen by more than half by the turn of the century. During 2003 it is expected that the total UK cod catch will be no more than 12,000 tonnes while the haddock catch will have fallen to around 30,000 tonnes. It must be emphasized that the cod and haddock problem has not been caused by the sectoral quota system but has occurred despite it. The reasons for the apparent collapse of the cod and haddock fishery could lie in fact that Governments have set quotas significantly higher than the scientific advice. Some have argued that the scientific advice itself has been flawed. It may even be that stock decline is related to some yet poorly understood environmental factors such as ambient sea temperatures or seal predation. Whatever the reason, the sectoral quota system has not by itself been the cause. The successful management of other fisheries under the sectoral quota system is an illustration of how well the UK programme of co-management has infact worked. Despite the current but specific problem with cod and haddock stocks, the UK has a system of co-management that has been successful because it has devolved real decision making and real management responsibilities to fishermen. It has also been successful because it has enabled fisheries dependent communities such as Shetland to establish community ownership of fish quotas.

Success in Fisheries Management?

Can these three innovations in fisheries management be regarded as successful? The Shetland sand eel management regime and the Shetland Regulating Order appear to have halted stock decline and have set in place the management framework that should ensure continued stock recovery. The situation regarding the UK sectoral quota system is less clear. While the pelagic fisheries are fairly stable, most demersal stocks continue to be overfished, with the cod and haddock stocks giving cause for real concern.

Insofar as fishermen have become directly involved in fisheries management, all three management systems can be described as a significant success. Indeed, the fact that the sectoral quota system still continues to command virtually unanimous support from demersal fishermen, despite the dramatic reductions in cod and haddock quotas, illustrates more than anything else how fishermen have bought into the fisheries management process.

The involvement of fishermen has gone further in Shetland where the wider community now has a significant role in all three management systems. The sand eel fishery and the Shetland Regulating Order, together with the community quota innovation, are all remarkable examples of community based fisheries management initiatives which are regarded as a huge success by the people directly involved - the fishermen and wider community of Shetland. There is perhaps no better criterion on which to assess the success or otherwise of how fisheries are managed.

Traditional Community-Based Management Systems (TCBMS) in Two Fishing Villages in East Godavari District, Andhra Pradesh, India

Venkatesh Salagrama[1]

Introduction

In India, fisheries development, reflected in emphasis on improving technological efficiency and increasing export growth, has accelerated over a period of half a century, radically changing the way fish are caught, processed and traded. This growth is accompanied by an increasing degradation of the natural environment and resources, due mainly to the unsustainable and increasingly non-viable, fishing and other coastal activities, leading to increasing vulnerability of a large majority of people, who are some of the poorest in the country.

As a majority of coastal capture fisheries have begun to show serious signs of overexploitation and declining productivity, the importance of effectively managing the coastal fisheries has assumed importance. It is realised that many of the management and control regimes being implemented elsewhere in the world (such as ITQs) cannot be easily applied in the Indian context, because of the largely informal nature of the fishing economy. Consequently, the current fisheries management strategies are based upon the premise that by controlling direct human exploitation of the resources, the natural environment can regenerate. Fisheries management has thus become synonymous with ways of exclusion of both traditional and modern users by means of seasonal bans, bans on fishing in specific areas and on catching certain fishes, and restrictions on using some fishing gears.

Besides the questionability of the effectiveness of such measures (which often ignore the influence of externalities – pollution, habitat destruction, upstream activities, and do not evaluate the relative impact of different activities) in improving the natural resources, such exclusion has a serious impact upon a majority of coastal population for whom fishing and allied activities are important livelihood activities. Lacking alternatives, many people are forced to ignore or get around the laws to pursue their occupations and the purpose of the management measures is

---

[1] Director, Integrated Coastal Management, Kakinada, Andhra Pradesh, India.

lost anyway. Efforts to implement the management measures have also been constrained because of:
- Multi-species nature of the fisheries;
- Informal nature of most fishing systems with nearly 80% of fishing carried out by traditional and motorised sectors (GOI, 2000: 128);
- Difficulties in managing a long coastline (over 8,000 km) covering 13 states and union territories;
- The need to balance management efforts with providing livelihoods and flow of foreign exchange earnings; and,
- Inadequacy of the formal institutional systems to implement management plans efficiently.

That fisheries is a state-subject, i.e., each maritime state in the country has the freedom to choose its own policies concerning fisheries, the Centre being largely confined to an advisory role, adds to the complexity of the issue. There also exist big gaps in the current understanding of the status of many coastal fishery resources, which constrains undertaking effective means of management.

On the other hand, there is evidence to indicate the existence of vibrant and sustainable modes of resource management and use patterns within the fishing communities. Contrary to the widespread belief – which informed and influenced the past development policies of inclusion and current management practices of exclusion – that most fisheries were open access, hence subject to the 'tragedy of the commons', there is good evidence to argue that "The important question is not whether or not there is a system for regulating access and withdrawal rights. Rather, the issue is about its structure or, if no system is found, about the conditions that prevented its development or caused it to disappear"[2]. This understanding is important in that it brings the primary stakeholders, i.e., the fishers, into the heart of the fisheries management. The remarkably extensive and in-depth knowledge of the fishers about their fishing environment and resources has been noted by many authors, and it can be surmised that this knowledge laid the foundation for the systems of management. The systems are often very location-specific, but in the coastal environment, which is characterised by an enormous amount of diversity, they cannot be expected to be otherwise.

The government's ignorance and/or disregard of the 'informal' systems meant that the knowledge and understanding accumulated over generations are not taken advantage of while designing the new systems. Moreover, by alienating the fishers from decision-making roles in the new management systems, the government virtually took the burden of implementation of the conservation and management plans upon itself, a huge task that it cannot fulfil on its own, with the result that most management plans remain confined to paper.

---

2 Bavinck, M. 2001: Marine Resource Management: Conflict and regulation in the fisheries of the Coromandel Coast, Sage: New Delhi.

This study is based upon three premises:
1. That there exists a close relationship between the features of a natural ecosystem (including, but not confined to, fisheries) and the social and economic structure of a fishing community in a particular area and that there is a need for policies to recognise and accommodate the resultant diversity characterising different systems in the fisheries management programmes.
2. That there exist traditional systems of management in several fishing communities, which reflect the diversity of the fishing conditions and systems, and evolved as a response to a need for managing the utilisation and distribution of the resources, and that their structure and functioning was based upon participative and consensual decision-making processes.
3. That an understanding of the customary systems of management can contribute to more effective and need-based policies concerning fisheries management and conservation of the natural resources in the coastal areas.

The study thus aims to understand the evolution of the community governance structures and their legal systems in the context of the features of the natural ecosystem that surrounded them. Taking Bavinck's definition of a legal system 'as consisting of a set of rules as well as the authority, or the organising entity responsible for its formulation and implementation', the study discusses the authority which the informal Caste Panchayat derived from its position as the supreme arbiter on all aspects of village life. The whole 'system', which refers to 'the harmonisation of, and the connection between, rules and authority', then, is more complex than being confined to decide who fishes when, where, and how, and understanding this provides some pointers to the useful aspects of TCBMS which are relevant to current fisheries management policies and programmes.

The location
The state of Andhra Pradesh (Ândhra Pradèsh) on the east coat of India has a rich diversity of coastal and marine ecosystems represented along its more than 900 km coastline. According to DOF (1999), the total number of fishers in the state in 1993 was over 870,000, of whom over 277,000 were men, and 260,000 women, and the rest were children. The coastal fishers, who numbered over 560 thousand, accounted for nearly 65% of the total fishers in the state.

The coastal area has been classified into north, central and southern zones by various authors, based on the geographical, physical and environmental features. Each of the zones is also represented by one major fishing caste.
1. The northern zone is characterised by open surf-beaten coasts, extending from Orissa border to Uppada, is dominated by fishers of the *Vadabalija (Vādabalija)* caste, interspersed with a smaller caste group called *Jalari (Jālari)*;
2. The shallow central zone, influenced by the large inflows of waters from the Rivers Godavari and Krishna, extending from Uppada to Nizampatnam, dominated by fishes of the *Palle (Pallé)* or *Agnikula Kshatriya* caste; and,

3. The southern zone, extending from Nizāmpatnam to Tamil Nadu border, once again characterised by open surf-beaten coasts, and by fishers of *Pattapu* caste group, who are closely related to the *Pattinavar* fishers from the neighbouring stat of Tamil Nādu.

Within each zone, there are many differences, and these distinctive features are reflected in the fishing systems, fish disposal and marketing systems, social and political organisation of the communities. The study concentrates on two fishing communities – one, an open-sea fishing community of *Vadabalija* caste in Uppada (Uppāda), and the other, an estuarine fishing community of *Palle* caste in Boddu Chinna Venkatāya Pālem (BCV Pālem, for short) – in East Godāvari district in the central zone.

A brief review of literature on TCBMS in Andhra Pradesh

Few studies have been attempted on the Traditional community-based management systems in the coastal communities of India, and this holds true of most studies dealing with maritime communities in South Asia (see Bavinck, 2001: 141). In spite of a hoary tradition of fishing in Andhra Pradesh, the existing literature does not do much justice to the subject. The few anthropological and ethnographic studies that exist provide only sketchy references to the customary tenurial systems in the fishing communities. Most studies dealing with the coastal fishing populations in Andhra Pradesh have tended to be either too broad-based with the TCBMS receiving only cursory attention, or too focused on certain issues, such as the vast body of work that the Bay of Bengal Programme of the FAO published.

Edgar Thurston (1909) may have been one of the first and certainly the most cited chronicler of the fishing castes, among others, in South India. Although his work is nearly a century old, and much of what he describes may have changed, it is still recognised to be the authority on South Indian castes. Curiously, the normally thorough district gazetteers of the coastal districts of Andhra Pradesh (Mackenzie, 1883; Francis, 1907; Hemingway, 1915) seem to ignore fishing communities and fishing systems more or less completely, which is perplexing particularly when set against the elaborate descriptions of the same provided by contemporary administrators like Thurston (1909) and Hornell (1915).

Suryanarayana (1977), Schömbucher (1986), Subbarao (1980), and Herrenschmidt (1989 & 2002) studied some aspects of organisation amongst the *Vadabalija* community, but the studies by Schömbucher and Herrenschmidt are in German and French respectively, making them inaccessible to the non-German and non-French readers. Schömbucher's study, the summary of which is available in English, appears to be comprehensive in its treatment of social and economic organisation of the communities, which makes the inaccessibility of the full study all the more glaring. Suryanarayana's description of the fishers of the northeast coastal Andhra Pradesh (1977) is one of the few studies that explain about the traditional systems of organisation amongst fishing communities, which also refers to the impact of the emergence of formal Panchayati Raj Institutions (PRIs) on the customary systems.

Some of the studies by the FAO-executed Bay of Bengal Programme do make a mention of the community and caste councils but with a few exceptions like Tietze (1986) bypass them largely. Tietze (1986) did a comprehensive ethnographic study of the *Vadabalija* community in southern Orissa, but the study pays scant attention to the customary use rights and management systems. Vivekanandan et.al, (1997, draft) touched upon the organisation of the fishing communities in Andhra Pradèsh, but their study having other concerns and aims as well, provides only a brief idea of the organisation of fishing and management.

The Census Survey Report of 1981, which studied Mofusbandar in the Srikakulam district in some detail, is a comprehensive account of the social and economic life of the village. The study is however silent on the fisheries management functions of the Caste Panchayat. While this could be considered as indicative of the relatively less importance of fisheries management to an open-sea based fishing village, the existence of shore-seines in the village (which require some form of tenurial arrangements), and evidence from elsewhere (including Uppada), indicates that the resource management aspects of the Caste Panchayats may have been overlooked in this otherwise exhaustive study. An update, prepared in 1981, does not throw much light on this aspect either.

The Tribal Cultural Research and Training Institute, in a monograph on the fishermen of Pudimadaka village in Visakhapatnam district (1965), touches upon the caste councils, but does not elaborate their structure and functions. A study by Razeq (1970) (cited by Tietze, 1986) is said to throw light on the fishing communities, but could not be obtained.

Amongst the fishing communities, within the four dominant castes, the *Palle* are the better educated. There has been a prolific output of papers, books and pamphlets by and on this community, dating back at least to the last quarter of the 19th Century. The published information by the *Palle* about themselves (for e.g., Samudra Rao, 1980) however tends to avoid the subject of fishing, appearing rather shamefaced about the occupation and is preoccupied with proving the links between the *Palle* (who style themselves 'Agnikula Kshatriya') and the Pallava kings of the Tamil country. The result is that these writings do not provide much enlightenment on the customary systems, or their evolution through different times. This lacuna is particularly acute because the *Palle*, by virtue of dominating the central zone of Andhra Pradesh, developed extensive tenurial arrangements governing the use of fisheries resources.

Sebastian Mathew (1991) does a comprehensive study of the territorial use rights in the fishing communities operating in the Pulicat Lake, bordering Andhra Pradesh and Tamil Nadu. The study is unique in the sense that it is – so far as could be gathered – one of its kind dealing with community-based fisheries management systems on the east coast of India.

It is thus possible to conclude that there are still big gaps in the understanding of the community-based legal systems and their effectiveness.

Methodology of the study

Field research was undertaken in two fishing villages. The study began with a survey to obtain an understanding of the community structures, including caste and occupational composition, followed by individual and group-based interactions. The key informants included elders in the village Panchayats, old fishers who were knowledgeable about TCBMS, as well as fishers of the present fishing generation. Women and other caste groups in the villages – who have reasons to feel marginalized by the TCBMS – were met with separately to obtain their perceptions of the systems. Quantitative and other information, where available, was obtained from secondary sources such as the village Panchayat, government agencies and NGOs working in the villages. Using this information, a basic picture of the Caste Panchayats is developed, which was validated and refined in subsequent group interactions[3].

One problem with a study of this nature is in determining what constituted a *traditional* system. It is recognised that the systems being described were not static and constantly evolved with the changing circumstances and conditions. There has been a speeding up of the deterioration of the TCBMS over the last decade, and in places like Uppada, a large part of the systems is already outdated, so it is difficult to determine the 'pristine' structure of TCBMS. On the other hand, a comparison with 'modern' systems is impossible in the absence of a construct that nearly approximated what has existed before. Lacking secondary information from literature, the study had to obtain a picture of the systems from primary sources. It was decided to use reminiscences to guide the reconstruction process, and validate them with triangulation, group discussions and where possible and appropriate, from written sources. The picture that is developed as a result may be considered as a composite of things from different times. The point to remember is that although Present Tense is used throughout to describe the management systems, it does not always mean that the systems are presented as they are today.

The nature of fishery in the two locations

The multi-species fisheries and the wide variety characterising the natural and physical conditions of fishing on the east coast of India have given rise to a diversity of fishing systems. Although the fishing systems in different parts of a coast bear close resemblances to one another superficially, a closer examination will reveal significant differences. However, this fact is often neglected in policymaking, hence needs to be made again in this context.

---

3   It has however to be mentioned that this study is confined to describing the basic contours of fisheries management by the Caste Panchayats, and is intended to inform a larger study dealing with the issues of legal pluralism and their impact upon the livelihoods of the coastal fishers.

A brief description of the villages

The two fishing communities are located 40 km apart from one another, but are characterised by very different natural ecosystems. The social and economic organisation of the fishing communities is largely determined by the fishing systems in the village. The following section is confined to describing the traditional fishing systems in the two locations, as adapted to fishing in the sea (in case of Uppada) and in the shallow backwaters (in BCV Palem).

Uppada

Uppada is a major fishing village in Andhra Pradesh with a population of 13 thousand people. It is an open, sea-facing village, with a long maritime history. The fishers are known for their adventurism and entrepreneurial skills. The village is a conglomeration of 12 hamlets, and each hamlet is a homogeneous entity in its caste composition[4]. Although Uppada is considered as a fishing village, other occupational groups – mainly artisanal textile weavers, basket weavers, traders, and suppliers of other services – constitute 30 percent of its population.

Boddu Chinna Venkataya Palem (BCV Palem)

BCV Palem is a fishing hamlet of about 4 thousand people in about 650 families, belonging almost entirely to *Palle* community. The hamlet is part of the larger village of Coringa, which had a rich maritime history until early 19th Century, but is now no more than an agricultural village, 20 km upstream from the sea. BCV Palem is located on one of the numerous creeks of the River Godavari, which opens into the sea approximately 20 km from the village. Large tracts of land used for agriculture and aquaculture surround the village. Although mainly a fishing caste, the *Palle* are also frequently involved in occupations other than fishing, for example, agriculture. The *Palle* also prefer to fish more in the creeks and backwaters. The BCV Palem fishers began fishing in the sea about 20 years ago because of competition in the creeks.

COMMON PROPERTY RESOURCES IN THE TWO VILLAGES

Fisheries management becomes important because the characterising feature of the capture fishing systems is their dependence upon the open access or common property resources. The availability of, and access to, the various open access/common property resources – fish, sea/backwaters, beach, mangroves, etc., determine the way fishing systems, and the legal systems for regulating their access to the resources, developed and functioned. Because of the open-sea based fishing in Uppada, the conditions there favoured a less detailed system of management (although it was more elaborate in terms of social organisation, which was

---

4   One of these hamlets has a community of *Palle* fishers, who provided a useful point of reference for comparison during the study.

built around fishing), while the community property nature of the resources in BCV Palem determined that the systems governing access to, and exploitation of, these resources were more rigorous.

There are differences in the varieties of fish caught in Uppada and BCV Palem. In Uppada, these consisted mostly of pelagic varieties, often with a distant provenance, are widely distributed, and include some straddling fish stocks as well. Important catches include sardines, anchovies, mackerels, croakers, ribbonfish, pomfrets, seer fish, and sharks, although the operations came to be focused on shrimp during the 1990s. The important catches of the BCV Palem consist of demersal species such as shrimp, crabs, mullets, eels and mudskippers – which are distributed over a relatively small area and their breeding grounds known to be located close to the fishing grounds. Whether the differences in origin and distribution of the fish have an impact on the mindset of the fishers or on the traditional fishing and management systems is a matter of conjecture, but there are indications that the creek fishers have a good knowledge of the breeding habits, times and places of the fishes they catch.

Casuarina and palm trees are the dominant plant species in Uppada. The traditional houses are built on walls constructed with casuarina and split Palmyra poles over which mud is plastered, and the conical roof – supported by casuarina and Palmyra poles – is thatched with leaves of the Palmyra tree. Palmyra leaves are also used for a host of purposes – making baskets for keeping fish (baskets of different sizes and designs for different species and varieties of fish) and for drawing water from wells, hand fans, mats to sit/sleep on, hats, playthings etc.

The sea beach was the most important natural asset that Uppada boasted, after fish and sea, and is one asset over which the community could proclaim ownership, and take management measures to control and regulate access. It was a focal point of most important activities in the village, whether related to fishing or not. However, located just north of the area where the River Godavari meets the Bay of Bengal, Uppada is constantly prone to erosion, and the village constantly shifts backwards as the sea encroaches upon the land, reducing the beach to an ever-narrowing strip of land. This has affected not only the fishing and related activities, but also taken away an important social space for the villagers. Erosion also led to the affluent fishers moving away from the beach, confining the poorest closer to the shore, increasing their vulnerability, as well as reducing their access to institutional support. The relocation of the villagers elsewhere has led to a realignment of informal structures and systems within the village, hastening the collapse of TCBMS in the village.

For BCV Palem, mangroves are the most important natural assets after the fish. The village has access to extensive tracts of mangrove forests. Houses in the village are constructed on a skeleton of mangrove poles, and mangrove reeds provide the roofing. Mangroves also had many other uses to the villagers: as firewood (both for domestic use and for sale), in fish processing, as cattle pens, for grazing cattle, for consumption and as medicine, for making bricks and shell lime, for colouring the cotton fishing nets, and for catching certain species like crabs and *Acetes* shrimp etc. The Caste Panchayats drew their authority from controlling access and use rights in the creeks passing through the mangroves, and derived a major portion of their income from them until early 1990s. The new opportunities for exploitation i.e., aquaculture, as well as the need for regulating access, in the form of government legislation, brought along a host of private and government interests into the area, which usurped or undermined the role of the TCBMS.

The fishing crafts

The traditional fishing crafts of Uppada were suited for operating from the open beaches. The wooden boat-catamarans – called '*Teppa*' – were used for fishing in the near-shore waters and the plank-built '*Masula*' stitched canoes, were mainly used in shore-seine operations[5]. The shape (streamlined body, narrow beam, upswept ends), and method of construction (logs or planks lashed/stitched together with ropes, and not nailed, making the boats flexible and easy to dismantle for carrying up the beach) of the boats reduced resistance to the surf, while making them agile to negotiate even rough sea conditions. The boats were propelled by rowing and paddling, as well as by taking advantage of sails where possible.

The third major fishing craft operating from Uppada – the '*Nava*' – is a link that connects Uppada with BCV Palem, being the main fishing craft in BCV Palem. The *Nava* is a ribbed flat-bottomed plank canoe, which is solidly built with metal rivets, and was not exactly an open-sea fishing craft, being more an extension of the riverboats. Because of their structure and construction, they are more suitable for fishing in calm waters, including the shallow near-shore waters, creeks and rivers, and they are mainly operated from sheltered beaches. The Uppada *Nava* was mainly used to fish in the sheltered Kakinada Bay to the south until motorisation gave it the power to venture into the sea[6].

In BCV Palem, the *Nava* is followed by '*Dhoni*', a smaller version of the *Nava* constructed to resemble a shoe, hence the name *Shoe-Dhoni*. It is suitable for fishing in the backwaters. The fishers of BCV Palem use the *Nava* for fishing in the Kakinada Bay, while the *Dhonis* are used to fish the creeks. Paddling and punting are main features of propulsion in order to negotiate the shallow waters in this area.

---

5   They were also used in marine fishing operations in some areas.
6   That Uppada is in the transition area between the open-sea fisheries of the north zone and the shallow water fisheries of the central zone accounts for the existence of the *Nava* in the village. It disappears completely beginning from the northern parts of Uppada for nearly 500 km until one encounters it again in Puri in Orissa state. There was a category of large *Navas* (over 28-feet in length) which operated from Uppada in the sea using sails, but these were more likely to be an adaptation from the central zone than a feature of the open-beach fishing systems, seeing that they were not seen elsewhere in the open beach fishing villages.

While the *Teppa* and the *Masula* were built locally in Uppada, the *Nava* was constructed by a group of specialist boat builders located in Tallarevu, a village about 2 km from BCV Palem in the Godavari delta area, another indicator of the riverine origin of the *Nava*.

In Uppada, there have been changes related to material and methods of boat building, fishing and the means of propulsion since 1980s, whereas in BCV Palem, there has been virtually no change either in boat construction or means of propulsion, as the shallow creeks offer little scope for either. This is a reason the TCBMS concerning fishing continue to exist in BCV Palem while they have largely lost relevance in Uppada.

The fishing gears

The type and location of operation of the fishing gears determines the control and management functions of TCBMS. The fishers of Uppada use a wide variety of gill-nets, trammel-nets, and hook-and-lines, all of which – by virtue of the depths at which they are used – necessitate being used from a boat, hence each fishing system came to be identified with one type of boat. The shallow waters of BCV Palem permitted operating several varieties of nets without the active use of a boat, and many nets independently made up a fishing system using the boat often only as a means of transport to the fishing grounds.

Thus, while the open sea fisheries of Uppada had two or three major fishing systems operating different varieties of nets at different times, the estuarine fisheries of BCV Palem were represented by a number of fishing systems all being in use simultaneously. Moreover, in Uppada, despite the existence of a variety of nets, all fishers used the same kind of gears at any given time depending on the predominant catch of the season. Competition was avoided because there was opportunity for everyone to fish in the sea, until more advanced systems began to compete for the same fishing grounds in the 1980s.

The long and wide sea beach provided scope for operation of shore-seines, which were an important fishing method of the *Vadabalija* in Uppada until a decade ago. Shore seining necessitated, and lent itself to, some form of management, and the most important fisheries-related function of the TCBMS pertained to fixing the rules of operation of shore-seines and resolving the frequent conflicts that this activity engendered.

In BCV Palem, each of the different fishing systems is adapted to a particular niche within the estuaries, but these were still contentious, and needed constant supervision by the Caste Panchayats. The fishing boundaries for different villages were customarily (and notionally) acknowledged and respected, although these became increasingly controversial in the 1990s.

The fishing gears in Uppada are mostly the floating and drifting variety, with no setting nets, because the depth of fishing is in the range of 10-20 fathoms. Fishing in the creeks and the Kakinada Bay seldom exceeded 5 fathoms, and the fishers of BCV Palem used a range of tidal nets, which were either fixed or set or dragged along the bottom of the creek.

The period of fishing

Fishing is confined to about 8 months in Uppada, with the sea being too rough or the catches too poor during the rest of the year. The fishers migrate during this period along with their boats to distant areas like Puri and Paradeep in Orissa, or to the southern towns within Andhra Pradesh. Seasonal migration, thus, is a way of life in Uppada, and its influence is reflected in a number of ways concerning the social and economic organisation of the village.

In contrast, fishing in BCV Palem is carried out throughout the year, except when there are floods in the river. The issue of migration seldom arises for the fishers. However, while the marine fishing caste of Uppada traditionally disdained any kind of occupational migration[7], this inhibition did not apply to the *Palle* fishers of BCV Palem whose women routinely worked in a variety of non-fishing related activities.

Fishing activities in BCV Palem are determined by the phase of the moon, which affects the tidal levels and fishing conditions. Consequently, for a few days every month, there can be no fishing. Different fishing craft find different periods in the lunar calendar appropriate for their activities; consequently, stress on the fishing grounds is reduced by natural means. In Uppada, where no such influence of the moon is felt on the sea, it is customary for the fishers themselves to refrain from fishing on every Thursday[8].

Fish processing

The long stretches of the beach near Uppada provided good opportunities for fish to be dried on the beaches. For salting, Palmyra stumps, after the trunk was felled for house building, were hollowed out and used as salting vats; larger catches were salted in pits dug into the sand. Different species of fish were processed in different ways – the duration of salting and drying, concentration of brine etc were specific to each individual fish, and even to different size groups of the same species. The finished product is packed in Palmyra baskets – which were made specifically for the purpose by a community of basket weavers living in the same village – and carried by bullock carts, and later on, by bus, to markets up to 60 km from the village.

In BCV Palem, the strongest proof of the ecosystem-based nature of fishing operations comes from a mode of fish processing that is unique to this region – fish smoking. Fish are smoked on bamboo platforms set up over hearths where mangrove wood and coconut husk (the two most abundant firewood materials in the delta areas) are burned continuously inside a closed hut, giving a product with a distinctive flavour, texture, taste. Good fish catches that could not be sold fresh readily because of inaccessibility of the delta villages, the fertile nature of the soil

---

7   This is changing – see Salagrama (2002).
8   As the number of converts to Christianity in the community mounted, they began to observe Sunday as the fishing holiday, so there are now two separate fishing holidays for the Hindus and the Christians respectively.

ruling out drying on large-scale, marshy nature of the beach and cramped village conditions, together with good availability of firewood material gave rise to this indigenous practice of preservation of fish. The markets for smoked fish, though lucrative, are confined to the Godavari delta, and the species smoked are particular estuarine varieties.

Summary

The fishing systems in these two villages developed based on the geographical, environmental and physical setting of each particular community, and thus, each fishing community – and the systems that it developed – often constituted an integral and indivisible part of the ecosystem that surrounded it both physically and metaphorically. It then follows that any changes – however small – that are wrought on the larger ecosystem have a cascading effect on the systems and processes in the fishing communities and vice versa.

This also suggests that uniform strategies of fisheries management across a whole area may not always work, and that local diversity must be somehow reflected in the policies, or in their implementation, in order to make them work. It makes the issue of fisheries management from a macro-level policy perspective more complex, but unless this diversity, and the way it came about, is properly understood, it will be very hard for fisheries management strategies to succeed.

It is also true that there are many similarities between the two systems – the most important and obvious one being their dependence upon a common, and fast depleting, resource. To rejuvenate the resource will mean tackling at a generic level a number of issues pertaining uniformly across a wider number of fishing communities. Issues such as the increasing influence of externalities and external players, uncertain/shifting tenurial systems, uncertainties and risk associated with traditional occupations, population growth, exposure to natural and human-induced 'shocks', and marginalisation and criminalisation of traditional livelihood systems will need to be considered as part of the larger fisheries management programmes for these programmes to be effective.

The nature of management

Thomson (1989) suggests that in natural fishing economies, the need for collective action arises from two reasons. First, fishermen cannot enforce a permanently viable individual claim on communal fishing grounds/territories because mobility and indivisibility of the biomass and the externalities in organising fishing operations make the costs of enforcing such rights exorbitant than the uncertain benefits from fishing itself. Secondly, the nature of cooperation in organising non-mechanised labour processes compels fishermen to stay together not only for enforcing their claims on communal fishing grounds but also for the viability of organising themselves in various production teams. A 'social contract' thus comes into existence as a response to the need for the collective to assert viable claims over the fishing grounds, as well as to ensure equitable access to all fishers in the collective.

The organising entity however is more than a production collective. Traditional community based management systems (TCBMS) encompass more than just fisheries management issues, and include the socio-cultural, religious, political, administrative and economic aspects of the community organisation. By legitimising the collective claims of the community, it not only performs its legal duties but also maintains the existing socioeconomic and political order in the village. Thus, a study of the TCBMS in the fishing communities serves two functions: firstly, it provides an understanding of the local fisheries management systems, and their relation to the communities' need for equitable usage of the resources. Secondly, it reveals the processes by which the informal structures of management could enforce their will, irrespective of whether the decisions pertained directly to fishing activities or not.

The fishing communities in Uppada and BCV Palem, which have been traditionally dependent on capture fishing, have well-developed community-based management systems. The traditional systems of management and control related to fisheries and fishing are more elaborate in BCV Palem where fishing activities are carried out by a number of fishing systems confined to the creeks and the backwaters.

In Uppada, such restrictions are not only difficult to enforce, but also superfluous to some degree. With fewer fishing systems using more or less similar fishing gears in a large and open area at sea, the Panchayats required fewer rules to govern them. On the contrary, the distribution of the economic product in social terms needed more attention, and the function of the management systems in Uppada related to ensuring social equity across the community[9]. Spatially confined systems in open-sea fisheries – like shore-seines – are subject to systems of management.

Bavinck (2001) finds a detailed rule system governing the access to the fishing grounds in the open sea communities on the Coromandel Coast, but concludes that, in the open sea fishing systems, 'fishing spaces are open to the entire population of artisanal fishermen (which largely, although not entirely, coincides with caste). Not only do all artisanal fishermen benefit from reciprocal access, but they also benefit equally. The similarity of fishing technology in the artisanal sector provides each participant with a more or less identical point of departure. In

---

9 There are indications that the open-sea based fishing systems were subject to restrictions on access to fishing grounds when the boats were non-motorised and the fishing grounds were close to the sea. It is interesting to note that claims of sea tenure are increasingly pressed in the open sea based fishing systems as well, where they had not existed before. The fishers of Uppada used to migrate to a number of places within Andhra Pradesh until recently, but are increasingly finding opposition to their activities from the local fishers, who object to outsiders fishing in 'their waters'.

conjunction, reciprocity and equality remain important clauses in the artisanal fishermen's rule of open access.'

The Caste Panchayat

The institution of caste provides the basis for codifying and ensuring effective enforcement of the customary laws in the form of '*Kula Kattadi*' or 'Caste Codes', the most severe punishment for transgression being ostracism from the caste. Collective management of common resources has been seen by many cultural anthropologists to be grounded in the fabric of Indian society, and it has been contended that the caste system – through the Caste Panchayats – plays a crucial role in its performance. There are examples from other sectors, like agriculture, livestock and forestry, which support this contention (Bavinck, 2001).

Uppada has one dominant fishing caste, the *Vadabalija*, while BCV Palem is a single (*Palle*) caste village, ensuring that their caste code could be imposed uniformly across all the fishing households in the village[10]. The following sections introduce the structure of the organising entity, its activities and the authority with which it implements its will upon the fishers. Obviously, the subject is too large to be delineated fully, and the salient features of the system that have relevance for fisheries management are only described here.

Structure of the Caste Panchayats

Panchayat literally refers to the system of village-level organisation with five elders ('*Pancha*' meaning five in Sanskrit), but it is also used to denote the general body of the caste group in the community. The general body is also referred to variously as *Kula Sabha* (caste council), *Grama Sabha* (village council), or *Kula* Panchayat (Caste Panchayat). The number of elders grew with raising populations, and in BCV Palem, where the number of elders is decided based upon the size of population, in a system of proportional representation, there are 11 elders now. The number of elders in Uppada, on the other hand, appears show a declining trend – some of the Petas have only one or two elders only. This may be a reflection on the decreasing importance of the Panchayat in Uppada.

At the basic level, the members of the fishing community constitute the general body of the Panchayat. The leaders are called *Peddalu* (elders; singular: *Pedda*) in Uppada[11], and *Pethandarlu* (managers; singular: *Pethandaru*) or *Peda Kapulu* (caretakers; singular: *Peda Kapu*) in BCV Palem. The designations provide a clue about

---

10 In villages where a dominant caste does not happen to be the fishers, it is still possible to see that the rules governing access and use rights in fisheries are largely controlled by the caste councils of the fishers, mainly because fishers and fishing often stood near the bottom of the caste and occupational hierarchies.
11 Although present tense is used throughout the description of these systems, the details pertaining to the caste panchayat are a mixture of what they were like until recently, and what they are now. Particularly in case of Uppada, the description of the systems is as they existed until early 1990s.

their functions: in Uppada, the elders play an active role in the social sphere of activities, whereas in BCV Palem, they have an important economic function, managing the access and use rights to fishing.

The members themselves are called simply '*sabhyulu*' (members) in Uppada, whereas in BCV Palem, they are called '*Paallu*' (meaning, shares; singular: *Paadu*), once again demonstrating the economic roots of organisation of the later system.

The powers of the elder are derived from their role as the custodians of the *Kula Kattadi*, or the Caste Code. This is an unwritten code of conduct for all community members, which is interpreted by the elders in dealing with the day-to-day issues of community life. This is also added to, modified or changed from time to time, depending on the conditions, based on consensus of a majority of members. Because their legitimacy depends upon ensuring the consensus of the majority and transparency in decision-making (or, at least a semblance of it), the fishers give the elders a respect that was not often accorded to the non-indigenous systems.

The fishing systems of BCV Palem that require organisation and management in terms of access and use rights to the fishing grounds have formed smaller groups, consisting of people of the same fishing orientation. These groups elect their own elders, who establish vertical linkages for the group with the Caste Panchayat at the village level, and horizontal linkages with groups operating similar fishing systems in other villages. This helps to overcome most of the fishing system-level problems at that level. The Caste Panchayat is obviously the final arbiter in the village, and all groups are subordinated to it.

Besides the *Peddalu*, the Panchayat in Uppada also employ two other functionaries, the *Sammita* – the village crier – and *Pillagadu* – the priest[12]. In BCV Palem, there is only *Sammita*, the priests not being a part of the structure.

Membership to Caste Panchayat

The systems in Uppada appear to have not only been egalitarian, but also strived to maintain that condition. This egalitarianism is found on the Coromandel Coast by Bavinck (2001), who wonders whether it is linked to the 'spirit of egalitarianism' that McGoodwin (1990) considers characteristic of fishing all over the world. In Uppada, membership in the fishing community – which forms the general body of the Caste Panchayat – is open to everyone born into the *Vadabalija* caste, with no restrictions on age, ownership of assets, social or economic status of a family, lineage or gender not being constraints for membership. All members in a family, including single women-headed families, are considered as members.

By ensuring the widest possible coverage, the Caste Panchayat ensured – being

---

12 Vivekanandan et al, citing Subbarao (1980), mention the existence of three positions of elders – the Kularaju, the Gramapilla and the Pilla, who were helped by a Yuva Raju – in *Vadabalija* communities, but, apart from the *Pillagadu*, the terminology does not seem to be used in the Godavari district. Suryanarayana (1977) too makes a mention of the existence of Kula Raju and Kula Pilla – while there are references to Kula pilla in Uppada, no mention of Kula Raju is found. It is possible that the Kula Raju system preceded – or existed alongside – the system described here.

more concerned with the social organisation of the village – that the social fabric of the village was not threatened by emergence of alternative power centres. The community-based nature of fishing occupations in Uppada would seem to have been a reason for the inclusive nature of its membership. Shore seines, boat launching and lifting, are all more or less dependent on the involvement of a large number of people in the activity. The fact that the Panchayat largely depends upon the donations of the members, and does not have an overt role in controlling fishing, meant that the more people that could join the Panchayat, the more income it earned. The predominance of small pelagics in the catches means that during certain parts of the year, the entire community has to work as one unit to be able to dispose off the fish properly.

Uppada being a multi-caste village, with a multiple occupational profile, and the *Vadabalija* being the dominant caste[13] in the village, its decisions often have influence over other sections of the community. People belonging to other castes are allowed to attend the Caste Panchayat meetings, but are ineligible for membership. In-migrants of *Vadabalija* caste are automatically considered members of the Panchayat.

In BCV Palem, on the other hand, the existence of use rights that are shared equally amongst the members has meant that there are more incentives to keep people *out* than in. Membership to the general body of the community is restricted to males of a 'productive' age. Women are excluded from membership, and so are children and old people. Single women-headed families cease to be members on the death of the male member of the family. That the Panchayat has access to other sources of income like CPRs means that it does not need to generate its finances from subscriptions alone. There are no group-based activities comparable to the shore-seines here. Non-*Palle* people in the village can attend the Panchayat meetings seeking redressal of their problems, but cannot become members. In-migrants from other villages belonging to *Palle* caste can become members by paying a fee as decided by the Panchayat, provided they obtain the confidence and approval of the people in the village. The 'entry fee' in a *Palle* village is quite hefty, but once paid, the new members share all privileges with everyone else.

Responsibilities of the members

All members are regarded as equal, carry equal rights and responsibilities, and are expected to follow the tenets of the Caste Code – *Kula Kattadi* – scrupulously. Members are supposed to regard Caste Panchayat as the supreme authority concerning the life and livelihoods in the village, and take all their complaints – domestic or professional – only to the Panchayat. Each member must attend the Panchayat meetings when called, and obey the resolutions and judgements passed by the elders. Members should contribute to the village funds, and depending on

---

13 For a more elaborate description of the 'dominant caste', see writings of MN Srinivas.

the elders' decisions, take up common fishing for community purposes, or refrain from fishing as and when required.

Rights of the members

Being a member of the Caste Panchayat confers the right to undertake fishing from a village on a fisher, who can take crew to operate his boat, or become a crewmember himself in others' boats. In Uppada, at least theoretically, one who is not a member of the caste body could still fish in the sea, this is impossible in case of BCV Palem, where individuals have access to communal fishing grounds only as members of the community, which is defined in terms of work, location, and shared cultural and spiritual practices.

The member-fishers in Uppada receive benefits like opportunities for diversification when sea fishing is not good, for e.g., work as labourers in local shore seine operations; depend on community assistance in fishing activities (launching and lifting boats to the shore; fish processing), or when boats or equipment are lost at sea by drifting or on land by thieving. Fish traders from the community can hope to receive fish on credit, which is denied to non-fishers. In distant areas where they migrated, the fishers of Uppada can rely upon their caste-men to extend a helping hand in setting themselves up. When a fisher loses his nets at sea, and is practically 'broke', his friends and neighbours contribute a piece of net each so that he can start working again[14]. Services of the village crier, barber, washers, etc are forthcoming to the members, who can also participate in celebration of festivals – '*Jatharalu*' and '*sambaraalu*' – along with the rest of the community.

In BCV Palem, being a member provided incentives, and not being so meant deprivation of basic entitlements. Only members of the Panchayat are allowed to fish in the creeks over which the village has sole use rights. Non-members can only fish in the Kakinada Bay, about 20 km away, which is considered an open access resource, and are only allowed passage through the creeks.

The members become part of a range of informal networks by virtue of their membership to the Panchayat, the different structures-within-structures provides opportunities for the members to take on issues of common concern in a collective manner. The moneys generated from renting out or otherwise making use of the village commons are shared amongst all members of the community. The village commons, unlike in Uppada, are rich sources of income to the Panchayat, and any surplus generated from leasing/renting out of these lands is distributed to members of the Panchayat. Members are also allowed to collect firewood, poles for house building, grass for feeding cattle or for thatching, etc., from the village's share of the mangrove forest. In the village, fresh drinking water being scarce, collection of tap water is restricted to members, and even wives of deceased members – who are excluded from membership – can obtain water only after the members and their families have their fill. Even at the time of cyclones and other

---

14 It is said that the nets thus contributed are carefully chosen to be the best among those available; when the fisher repays the loan in due course, many refuse to accept it.

natural calamities, the damages sustained by the members are given priority over non-members when submitting estimates of losses to government for rehabilitation[15].

Considering the benefits, which cover important aspects of their life and livelihoods, people have little difficulty in willingly and implicitly followed the caste code.

The institution of elders

Contrary to external perceptions, the system of elders is not always an elitist, exploitative and parasitical institution – although these features have come to figure increasingly prominently as the traditional systems collided with, and adapted to, the more formal systems. The election and functioning of the elders was strongly rooted in the democratic tradition. To become an elder, one must be considered intelligent and knowledgeable about the caste code. He should have good oratorical and diplomatic skills, as well as to accommodate different points of view. He should be acceptable to all sections of the community, with a reputation for being impartial. Older people are preferred over the young. In Uppada, although women are not exactly encouraged to be a *Pedda*, it is said that a woman had been elected to the post at least a few times in the past.

In Uppada, a *Pedda* is elected for life. The election of new elders is an open process, and all members must attend the Panchayat meeting on the day of selection without fail. In Uppada, the selection of a *Pedda* can be a short process, often a foregone conclusion, but subject to discussion, debate and ratification by the majority of fishers. There are times when the process drags on for a long time amid heated debates. Once elected, however, a *Pedda* is accepted by all members and given proper respect. Members can decide to impeach a *Pedda*, were he to lose their confidence, or to act in a manner contrary to the best interests of the community, or be seen to be acting in a parochial or biased manner.

In the representational system of election that exists in BCV Palem, the elders are chosen once every two years, although the same people could go on being selected repeatedly. This, it is said, is an effective check on the *Pethandarlu* misusing their access to village funds. Because a Pethandaru needs to keep his flock together, he cannot afford to boss over them. Any member who can prove to have a sizeable number of *Paallu* (i.e., members) behind him can become a *Pethandaru*[16]. The selection process ensures due representation to all geographical areas and lineages within the village. The aspirants are occasionally asked to prove their 'strength' in the village by giving the full details of their supporters, a sort of 'open ballot', which could potentially favour the more powerful and affluent

---

15 Obviously, the plight of non-members who were excluded from the system was extremely miserable; and when it is considered that the sufferers were widows, children and old aged people, who were cast off for no fault of their own, one can begin to see the enormity of injustice and callousness that are the other face of the caste system.

16 The bigger lineages in the community had thus a better chance of election than others.

over the others, but as the villagers argue, the dependence is mutual.

That the process of election – even though often a foregone conclusion – takes hours of debate, that some of the claimants do get rejected by the general body and in some cases, some incumbent *Pethandarlu* are removed in favour of others by a majority vote, shows more than a notional form of democracy at work.

Respect to the elders

The persons of elders must be given respect, and those who insult them will be reprimanded by the Panchayat. In BCV Palem, members are expected to show respect to the *Pethandarlu* to the extent of not sitting next to them in a Panchayat meeting, and needing their permission before speaking. There is scope for every member to speak at the assembly, voice his dissent, or argue, and if still unsatisfied, to take his case to the supra-village council (see below), but this can be done within acceptable limits of behaviour, and those who disobey, are punished according to the Caste Code. That the Caste Panchayats revel in endless debates is undeniable, going by the fact that even trivial issues can take hours of discussion before any decision is made. People, who repeatedly showed disrespect to the elders, or to their decisions, are penalised with increasing degrees of intensity, which in extreme cases leads to ostracism[17] from the community.

In case of disputes related to fishing, utmost priority is accorded to solving them before any other business is undertaken. This is because, in case of BCV Palem, whenever there is a dispute involving a particular fishing system, all such fishing systems in the village must refrain from fishing for as long as the dispute is not settled to the satisfaction of everyone. Although, generally, the fishing system-level group leaders tried to resolve the issue, matters sometimes are more complicated and necessitated involving the Panchayat. The other side of this is that when the fishers of a village refrain from exercising their traditional rights to a fishing ground, those from other villages are free to make use of them during the period.

THE AUTHORITY OF THE CASTE PANCHAYAT TO IMPLEMENT ITS WILL

Compliance with the system

The Panchayat ensures compliance with its directives through an elaborate and complex system of physical or psychological coercion, the means for which are at its disposal as the supreme legal and political authority in the village. It is built upon a system of incentives for obedience and punishments for transgression. The preceding section described some of the privileges that membership to the Caste Panchayat carries. The punishment for transgressing the caste codes or for not obeying the Panchayat's decisions ranges for imposing a fine to ostracism from the community, the latter being the most effective means of stripping a person of all forms of social security.

---

17 Veli – ostracism from the village – is banned by the Government of India, and is no longer applied in either village.

This may look like the Caste Panchayat is no more than an informal, but efficient, totalitarian regime. The significant difference is that the Panchayat derived its authority from its participatory nature of functioning. By making each decision a public affair, transparency and accountability are ensured in the system. Its legitimacy comes from the trust that the fishers have in its impartial nature – the Panchayat elders know that if this trust is broken, people can easily take their complaints elsewhere, as happened in the case of Uppada, where the Police have taken on many of the roles of the traditional Caste Panchayat these days.

The Panchayat also receives sanction to impose its will by virtue of its role as the custodian of the religious aspects of community life. In this, its role is similar to that of the Church in the southern states of Kerala and Tamil Nadu (see Thomson, 1989), and the importance of the spiritual angle in ensuring compliance with the directives of the Panchayat cannot be overlooked. If, in villages like BCV Palem, the Caste Panchayats continue to exert considerable influence on the course of events, it is also because of their identification with the religion.

Enforcement of Panchayat decisions

The issues brought before the Panchayat are deliberated at length in the caste council meeting before any decisions are arrived at. The disputants having been sworn in, and expressed their faith in the justice of the system, which is done in the public domain, cannot but bow to the judgement of the Elders. In BCV Palem, in cases where either the complainant or the respondent is a *close* relation to a particular *Pethandaru*, the *Pethandaru* is automatically disqualified from sitting in judgement over that particular case.

Most of the punishments in the villages seem to have been in the form of fines. Desertions, divorces, elopements, or drunken brawls, all are met with a fine, which is a major source of revenue for the Panchayat (and more importantly, to the elders). The Panchayat in BCV Palem also has the right to sell or auction off a guilty person's properties to pay off his '*Tappulu*'.

That fine is the most common form of penalisation in both villages indicates the need for the Panchayat to generate funds for their own operational costs. This could be argued to be a weakness of the system, because there is always potential for concerns other than strict administration of justice to play a role when dealing with a case, and might give the members the feeling that transgression of most kinds can be 'bought off' by paying a fine. It also gives an edge to the more affluent people in the community in terms of their ability to buy their way out of a problem. On the other hand, the community informants argued, surplus being always in the short supply in a subsistence-economy like theirs, *Tappu* – either in cash or kind – is a major deterrent for most people to contemplate any 'anti-social' activities. In traditional rural communities, most members of the community are at a similar economic level, so there is little scope for the 'moneyed' people to buy justice. It is only after fishing became a cash-based economy that paying a fine has come to be regarded as an easy option to overcome the community restrictions.

The transgressors of the community law were let off with warnings for the first few times, but when all other avenues were closed, would be ostracised by the consent of all elders in a general body meeting. The ostracism – '*Veli*'[18] – involved withdrawal of all benefits that a member enjoyed (see above), plus imposing many additional hardships. The Panchayats in the neighbouring villages too would be informed of the '*veli*' to ensure that the family obtained no assistance from elsewhere[19].

The ostracised person could repent in due course, appear before the Panchayat asking for a just punishment, and comply with the new punishment without demur, in which case he would become a full member of the community once again. Occasionally, people would leave the village for good, and migrate elsewhere. *Veli* was not intended as a permanent condition, unless the member concerned was too recalcitrant to get back into the fold – it was a very effective means to bring the errant individuals to follow the caste code. When a complainant had reasons to feel that justice was not done, he could petition the supra-village Caste Panchayat.

Resources available to the Panchayat

In both villages, all members are expected to contribute to the common fund set up for holding festivals and other programmes. The amount to be contributed by each member is decided by the elders. The Caste Panchayats generate funds from *Ummadi Veta* (community fishing) and collections from individual members – '*Ethubadi*' – in the form of donations for specific programmes. Besides these, a part of the penalties collected from disputing parties are deposited in the common funds.

The elders have the right to declare '*Ummadi Veta*' (community fishing) at any time of their choosing, although some advance notice is generally required. On the day set for *Ummadi Veta*, all members of the Panchayat are obliged to go fishing, and the truants will be made to pay an equivalent of the average income per fisher on that day.

In BCV Palem, the village commons are productive, and yield substantial revenues. The common lands are rented out for cultivation of paddy and groundnut, and later for aquaculture. The firewood from the mangroves was also auctioned off. The hamlet, along with its twin hamlet BPV Palem, has joint use rights to the drain, which separates the two hamlets, and this is auctioned off for

---

18 'Outcaste' is an apt word for one who is ostracised from the community: it literally meant losing one's caste for the duration the family was in disgrace. It also indicates the power that could be exerted by invoking caste as the threat factor.
19 At a time when matchboxes had not yet made their entry into the villages, generating fire for cooking purposes at home was always a major task, which involved constant borrowing of faggots from one another. The most serious handicap that an outcaste family faced, after denial of water, is this sharing of fire. This gave rise to a Telugu phrase meaning denial of fire and water as a general threat: "You will not get fire and water in this village!"

leasing annual fishing rights. The village freshwater tank is also auctioned off to fish culturists. Besides, the rights to sell liquor in the village and the more dubious rights to card games etc. also yield sizeable amounts of money. The village Panchayat also generates considerable revenue from the temples in the form of offerings by devotees.

Another important source of income to the Panchayat is the revenue generated from *'Tappulu'* (fines). In BCV Palem, the income from *Tappulu* takes care of the elders' time, because they share the amount equally amongst themselves. In Uppada, this may have been the most important source of income to the Panchayat, which lacks any other productive sources. Membership fee collected from new arrivals in BCV Palem – though substantial– does not amount to much, as new membership is given out grudgingly.

Activities undertaken with the resources of the Panchayat include contributions to the building of schools, latrines, roads, dykes, freshwater supplies and other infrastructure facilities in the villages. The Panchayat also pay for the services of people such as washers, barbers, *Sammita*, keepers of cremation grounds, nurses, priests etc.

In BCV Palem, all surpluses generated after payment of salaries and expenses for *Sambaraalu*, and upkeep of the village temple and other infrastructure is shared among the members of the community. Each *Pethandar* is given a certain lump sum, which he is expected to share equally with his supporters.

The Panchayat meetings

The meetings always take place in the Ram temple in both villages, and begin in the morning. The caste council – '*Kula Sabha*' (also called Panchayat)– meets under the direction and leadership of the elders. All community members are expected to attend the caste council meetings. Normally, a meeting is called only when there is something important to be discussed or decided upon, and always on days when there is no fishing. The Panchayat has the right to call anyone in the village before them. Villagers from the other communities can also be made to appear before the Panchayat after informing the *Pethandarlu* of the other community.

Regional Panchayats Council

In Uppada, the *Peddalu* of different *Petas* come together to form the regional Panchayats council, which meets frequently to discuss the common issues and concerns pertaining to all *Petas*, as well as to resolve the supra-*Peta* quarrels and disputes. When the disputes between two *Petas* cannot be solved amicably between themselves, even when the mediation of a third party is used, the dispute will then be brought before this supra council, which has overriding authority over all the community issues, and has the final word on all disputes. There are close linkages between the Panchayat in Uppada and all neighbouring villages, and the *Peddalu* of one village are constantly invited to attend a dispute resolution meeting in another.

In BCV Palem, there is a hierarchy of councils, starting from the one at the fishing unit level. When there is a dispute about sharing between fishers of the same village, the elders of that particular council will hear both sides and take a decision. When the problem cannot be resolved by the existing group of elders, they select more people to sit in the discussions to decide the case. When the dispute is not resolved even then, it is taken to the village Panchayat. Of the 11 *Pethandarlu* in BCV Palem, at least 8 must be present to give a hearing and resolve such conflict. In cases where other villages are involved, the *Pethandarlu* of the concerned villages are also roped in, and in extreme cases, the elders from an additional four villages will be called over. In all, about 60 people will then sit in judgement over the issue.

Key management features of the Caste Panchayat

Bavinck (2001) notes that fishing panchayats on the Coromandel Coast always rallied to restrict the usage of a fishing appliance, which was felt to harm the group. This harm could be to the fish stock, to the majority style of functioning, and to the community as a social entity. In any one instance, there is an intermingling of these themes. This perception of harm is applicable equally to the fishers of the two villages under study. The functions of the Caste Panchayats are built around these three issues, besides the more important one of ensuring continued recognition of the fishing rights to a particular territory.

The main fisheries management functions of the Caste Panchayat involved can thus be categorised into:
1. Assertion of rights over fishing areas
2. Balancing fishing activities with resource capacity
3. Establishment of rules of access for equitable distribution of fishing rights
4. Establishment of systems of governance that help to maintain the social integrity of the village

Assertion of rights over fishing areas
Caste Panchayats play a vital role in terms of asserting the community's claim on a fishing ground. Firstly, they ensure that a single caste group – *Vadabalija* in Uppada and *Palle* in BCV Palem – has monopoly over the use of any specific resource from a given locale (and thereby curtail the influx of people into fishing). Secondly, they guarantee that the collective claims of tenure of the community over a particular territory are recognised at the supra-village level, by entering into mutual agreements for recognition and safeguarding of fishing rights. To achieve this, a number of regional alliances are forged with different villages, much like in the international arena, and these ties are frequently reinforced. The hierarchy of caste councils provided efficient systems to settle disputes.

Between villages, there is constant friction in deciding the extent of use rights in the creeks, there are also conflicts between fishers while fishing in the creeks and between different fishing systems belonging to different villages, and these would all involve the intervention of the Panchayats of different villages to sit together for resolving the disputes. An important function of the elders in BCV Palem is as spokespersons for the fishers of the community in inter-village and intra-village conflicts. After the introduction of mechanised boats, which often compete for the same fishing grounds in the Kakinada Bay as the BCV Palem *Navas*, and overrun their nets with unfailing regularity, the role of *Pethandarlu* in taking up these issues with the mechanised boat associations has acquired increased importance.

It can be surmised that assured access to a common property resource in BCV Palem has given rise to processes of exclusion from membership, whereas the lack of – or irrelevance of – such rights in the open sea fishing systems allows a more generous distribution of membership. *De facto* ownership of a fishing ground ensures that the fishers have a stake in its wellbeing, and try to reduce pressure on it from other sources, while exploiting it judiciously themselves.

The closed nature of membership to the Caste Panchayat as well as to the specific fishing system-groups provided an insurance against competition from their fellow-caste members both within and outside the village. This is an important incentive because it provided protection to the livelihoods of the people. The fee for membership to the new entrants in a village was set so high that few people moved out of their villages.

Within the village, new entrants into *Vala Kattu* are allowed to join the group on payment of a nominal sum on a particular day – 15th January of every year – and no one can enter the system subsequently, until the next year. The entry used to be restricted to the number of people leaving the system during the year, but over time, the total number of units operating from the village has come down very significantly, hence the restriction has been lifted.

Balancing fishing activities with resource capacity

The resource management function of the Caste Panchayat is not often an explicit aim of the activities. There are no effort control methods in either location of study, and practices such as shrimp seed collection, though widely agreed to lead to wanton destruction of the resources, have not been subject to a ban by the Panchayats, which point to the Government as the agency responsible for such measures. However, there are indications that traditionally, the fishers as well as the Caste Panchayat were mindful of the need for ensuring sustainability of operations, and took measures to do so. Considering that their access to fishing grounds is limited to particular areas, it is possible to argue that the Panchayats – and the fishers themselves – do consciously try to manage them sustainably. Long-term sustainability of the resources is an important concern for the communities, who, after all, tend to lose more than anyone else if the resources are overexploited and declined.

The information from BCV Palem supports Mathew (1990)'s contention that, 'Considering the distribution of the fishery resources, the limitations of the fishing ground and the preponderance of fishermen around the [Pulicat] Lake, it is clear that the *padu* system has contributed to the sustainability of the lagoon fishery. In spite of the fact that conservation of the resources is not the principal aim of the *padu* system, the control over access-rights – limiting them only to the members of the *padu* system – perhaps has directly contributed towards preventing a collapse of the lagoon fishery.' As with most of the functions of the Panchayat discussed in the foregoing sections, the Caste Panchayats chose to relinquish this function only in later times both because of increasing external pressures and the government's takeover of the management function.

The restrictions in entry into the systems not only restricted the entry of outsiders – even of the same caste – into the communal waters, but also necessitated measures for efficient and sustainable utilisation of those waters. It allocated use rights for different fishing systems according to their needs, while taking into the consideration the capacity of the natural system to withstand the pressure that the systems exert upon it. Where the available area of fishing cannot accommodate all the fishing systems in existence in the village, fishing is allowed on a rotation basis. In *Ethudu Kapulu*, there are 38 units in the village, while only 27 units can be operated at any time in the available fishing grounds. The community resolved that only 27 units would be operated during the *Padunu* (peak fishing period in a fortnight, which is based upon the moon phases), while the other units wait on the shore for their turn in the next *Padunu*.

It is said that in a few villages on the Godavari Delta, fishing rights for *Vala Kattu* rested with particular families of fishers, and only those families could operate the nets. The shares for fishing were confined to a fixed number, at the rate of one share per family, irrespective of the active members in the family. The share is a permanent property, and a family could rent their share to others, when it cannot undertake fishing itself for any reason. The share is hereditary, and can be sold like any other asset. As a fisheries management measure, this works well in that it restricts entry for new people.

In Uppada, the Peddalu did attempt to dissuade the fishers from collecting shrimp-seed, mindful of the destruction it caused to the resources as a whole, but could not enforce their will because it concerned the livelihoods of a large majority of the people.

Fishing holidays and bans

Fishing holidays are another important feature of fisheries management in both the villages. The elders in Uppada have the responsibility of ensuring that all fishers kept off fishing for one day in the week. Obviously, observing a weekly holiday – ostensibly to give time to the fishers to rest and recuperate, as well as to meet their land-based obligations – and strictly enforcing the rules concerning the weekly holiday, show that it is as good a fisheries management measure as any. During lean fishing periods, *Peddalu* could allow fishing on the non-fishing days also to enable the families earn something for their subsistence needs. In BCV Palem, the dependence upon moon phases ensures that the fishers must stay on shore during parts of the month.

Fishing holidays are related to natural and human factors. Seasonal fluctuations in the fishing conditions and unfavourable weather (shocks including cyclones, floods, summer heat etc.) are the natural factors responsible for a majority of fishing holidays. The human factors – at the community level – that have relevance in terms of fishing holidays are festivals and seasonal bans, imposed by the community upon the fishers.

Festivals account for about 25-30 fishing holidays – stretching from one day to 10 days at one time – over a year. The Panchayat restricts fishing during the festival periods on pain of severe penalties. The fishing system groups in BCV Palem also have their own rituals and *sambaraalu* – for instance, during the month of *Margasira* (December-January) period, when there is a general outbreak of jellyfish in the creeks and the creek-mouths, the *Ethudu Kapus*' association refrains from fishing completely, and conducts *sambaraalu* to pacify the gods and improve the conditions.

The Panchayats declared a ban on fishing – '*Kattadi*' – for extended periods particularly in times of festivals, but also when extraordinary conditions necessitated a calling of the Panchayat meeting on a fishing day. The announcement comes generally at short notice, depending on the emergency of the situation, and the fishers are expected without fail to attend the village Panchayat meeting where a solution is thrashed out collectively. During periods when the sea was rough and cyclones were forecast, the Panchayat in Uppada declared *Kattadi* to keep the reckless fishers from venturing into the sea. All members were obliged to respect *Kattadi* and refrain from fishing on the days when it was in force. The non-observance of fishing holidays or transgression of the Panchayat *Kattadi* was met with a '*Tappu*'. The sudden declarations of fishing holidays to attend village meetings can account for up to 10-12 days in a year.

While more work still needs to be done, it is certain that the fishing holidays (that often come in the middle of a good fishing season) have a management function, which is implicit in the way they are implemented by the Panchayat. When it is remembered that there is a clear link between the features of the natural environment and the social and economic structure of the community, it begins to make sense to understand the village festivals as a means of fisheries management as well. The extensive knowledge that the fishers have of the biology and the breeding habits of different local species may be used to establish a correlation between their festivals and the breeding habits of the fish in their area.

Preventing resource rent leading to excessive/redundant investment in capacity
By ensuring that the fishers in the village remain confined to a particular area of fishing, and that their operations even within those fishing grounds needed to be sanctioned by the Caste Panchayat, the opportunities for the fishers to reinvest any surplus in more fishing units were curtailed. While it is not uncommon for an open-sea based fisherman in Uppada to own a number of boats, it is seldom that a fisherman in BCV Palem owned more than one boat. Even in Uppada, as long as

the fisheries remained non-motorised, the households were mainly content with one boat. It is also possible that by conferring a ritual status on the well-off families, the Caste Panchayat managed to extract any surplus from them for religious purposes, or for redistribution to the poor and destitute in the village. In management terms, this prevented over-exploitation of the resource, while maintaining social harmony.

Establishment of rules of access for equitable distribution of fishing rights
The one area where territorial claims become important in the fishing activities of Uppada, and necessitate an intervention by the Panchayat, is in the case of shore-seines. The operation of shore seines is a different matter from sea fishing for two reasons: one, until late-1980s, it was the most important fishing method in the village, and the catches from the 12 nets might have accounted for up to 80 per cent of the total landings. Secondly, being land-based, and necessitating some space for their operation, shore seines involved the issue of use rights – at both intra- and inter-village levels.

Consequently, the systems governing the usage of shore seines are very elaborate. Depending on the space available to each *Peta*, the coastline is divided into a number of lots, and shore seines are allowed to be used in different lots on a rotation basis in such a way that every shore seine has an opportunity to fish in all the lots at regular intervals. At a particularly lucrative lot, once the team whose turn it is to fish in the lot has finished its operations, others are allowed to run their nets for the second and third cycles. It is said that during the period when shore seining was a flourishing occupation, new entrants were actively discouraged, because the length of coast line over which the village could assert its rights limited the number of nets that could be operated.

While there are few rules governing the access to fishing grounds in Uppada, there however are controls to ensure that the fishing systems did not end up harming one another. When a boat overruns another's nets, it is fined. The fishers must also respect the fishing grounds where other fishers have traditionally fished. For instance, the Uppada fishers cannot operate near river mouths, where local (*Palle*) fishers might be operating their nets. The visiting fishers can ask for permission before shooting their nets, and the permission might involve sharing a part of the catch with the local group. In cases where the fishers violate this and fish 'illegally' in what is considered to be the *Palle's* waters, their own Panchayat takes them to task (on receiving complaint from the other village), which might take the form of an admonition the first time, but a stiff fine the second time around.

In BCV Palem, the three major estuarine fishing systems – *Ethudu Kapulu* (dip net), *Vala Kattu* (stake net), and *Pakkidevi Vala* (estuarine drag net) – are subject to tenure arrangements, which are extremely complex and detailed. Where the number of fishing units exceeds the space available for fishing, the systems operate by rotation. The multiple day fishing trips in Kakinada Bay too are timed in such a way that at any given time, a third of the boats are ashore.

The accommodation of different fishing systems works across villages also. BCV Palem has five boundaries within which it can employ *Vala Kattu* fishing, which it shares with BPV Palem fishers. Because of the differences in productivity of three of the boundary locations, they are rotated between the two villages annually. The other two locations are much bigger and more productive; hence, the two villages share them on a daily basis. Because of the spatial distribution of these locations, BPV Palem gets to use to two fishing areas in each location for every fishing area that BCV Palem gets to use. Thus, on a given fishing day, it is common to see two nets belonging to BPV Palem laid one behind the other, followed by one BCV Palem net, which in turn is followed by two BPV Palem nets, and so on. Within BCV Palem, there are 23 units involved in *Vala Kattu*, and they decide who fishes where within the five locations by a simple system of lottery.

The rule systems extend to the quality of exploitation as well. In *Vala Kattu*, which is a tidal stake net, the net is spread across the main creek during high tide period, and the fish are caught as the water flows backwards at low tide. A number of nets are in operation at intervals along the main creeks, which means that the nets spread out upstream are more likely to catch many fish compared to those behind. However, the nets to the back also get good catches of fish, which reach the creek through the numerous branch canals on either side of the creek. It is often possible for the fishers upstream to put obstructions across the branch canals at the place of their origin, in order that fish cannot either escape through the branch canals, and all those that do will come back into their nets. It is said that one net could catch 10 times as many fish as it would otherwise get by this means, so naturally there is a lot of temptation to obstruct the branch canals. However, the fishing group Panchayat frowns upon such practices as depriving the other fishers of an opportunity to catch fish, and considers them seriously, often penalising the culprits heavily.

Conflict resolution between different fishing systems is another important function of the Caste Panchayat. The most frequent disputes arise between the *Pakkidevi Vala* (estuarine drag net) and the *Vala Kattu* fishers, who often compete for the same fishing grounds. While *Pakkidevi Vala* is dragged across the creek, the *Vala Kattu* is a fixed fishing system with the result that one is seen to be harming the interests of the other, leading to violent confrontations.

ESTABLISHMENT OF SYSTEMS OF GOVERNANCE TO MAINTAIN THE SOCIAL SYSTEM

The social functions of the Panchayat are equally important as those related to fishing.

Community dispute resolution

Though the villages were largely homogeneous in social and economic terms, they were by no means peaceful, and there were always disputes – most of them petty, which, according to an informant, were 'an important entertainment in the village in the absence of cinemas and other diversions'. The foremost activity of the village Panchayat was settling the village disputes. Disputes related to land ownership, sharing family property, loan repayments, ill treatment of aged people by their children, marital infidelities, desertions, divorces, elopements, and drunken brawls – all were grist to the Panchayat mill. Most common offences were punished with a fine.

Significantly, the most important and time-consuming task of the Uppada Panchayat is settling disputes concerning transactions between boat owners and their crew. When a fisher dies, generally the owner of the boat on which he worked pays some compensation to his wife, but when this does not happen, the Panchayat steps in to help the hapless woman.

Reconciliation was the most commonly arrived at solution for many disputes brought before the village Panchayat (this is particularly applicable in case of marital disputes). However, this did not extend to all disputes – while some offenders were let off with a mild to strong censure, the more serious offences received punishments varying in degree from a fine – '*Tappu*' – to flogging.

Administrative functions

One important function of the elders was also to maintain law and order in the village. They had the authority to conduct death inquiries and ascertain causes of death, which, once established, were deemed final. The Caste Panchayat assists a fisher at the personal and professional level, and helps to keep the cost of litigation low. Legal assistance from the community also reduces the need for interacting with the Police and other law enforcement agencies, which is not only costly, but also time-consuming. The relatively inaccessible nature of their habitations, the sparse nature of their interactions with the outer world engendered among the fishers a sense of suspicion and weariness towards outsiders.

Going to the Police or external legal bodies is frowned upon by the Panchayat, and actively discouraged. When people sidestep the Panchayat in this matter, the Caste Code is invoked and an appropriate punishment is meted out, because the police are an alternative power centre, with more legitimacy and powers, and who can threaten the informal structures of governance. That the Panchayat decisions are relatively fast and binding on all members ensures a confidence in the spirit of law in the village. In fact, when disputes involving the community members are taken to the Police, the Police in turn call for the elders' assistance in settling the issue.

The elders thus acted as a link between the community and the external world, taking the community grievances to the government functionaries, who generally resided elsewhere, and acting as the first point of contact for anyone visiting the village from outside. The selection of beneficiaries for government development programmes was left to the Panchayat, on the unwritten understanding that any problems that arose subsequently were for the elders to solve.

Religious and social functions

The festivals are of two kinds: there are the Hindu festivals like Sri Rama *Navami*, which are celebrated uniformly across the area, although with different levels of intensity (the *Vadabalija* spend a week or more celebrating the festival, while the *Palle* spend only two days). The other festivals are more localised in nature, related to the village gods, of whom each village generally has its own set. These festivals take much longer, and the villagers refrain from fishing throughout the period. Besides collecting funds for these celebrations with the proceeds of '*Ummadi Veta*', the elders also form management committees for organisation and maintaining law and order during the period of the festivals.

Ummadi Veta can also be organised either to pay for some expenses related to a communal issue, to assist a destitute family, or to provide for the visit of an important dignitary to the area. As suggested by some authors (see for e.g., Thomson, 1989), the fact that *Ummadi Veta* always corresponded with good fishing seasons could mean that this was a way of siphoning off whatever little surplus there was to be had from the fishing by the political-legal authority for creating common goods and services. This prevented the degeneracy of common property and the community, by checking the growth of private property in the instruments of labour. These inbuilt mechanisms delayed the process towards greater 'privatisation' and enabled traditional communities to maintain an overall environmental and socio-economic harmony in the village.

The most outstanding example of the social function of the elders in Uppada comes into play at the time of weddings. It is the elders who are involved in the finalising the wedding negotiations, who consecrated the marriage with their presence, and gave it a stamp of legality that comes in handy when the marriages break up subsequently for any reason.

THE NATURE OF SUCCESS

The success of the TCBMS can be summarised to be a result of:

Their ability to understand and build upon the specific conditions of natural and social environment in an area

The management systems in the two locations have grown differently keeping the conditions of fishing and the organisation of the fishing activities. While those in Uppada concentrated on the social organisation part of the activities, the systems in BCV Palem, on the other hand, took on a more economic function, which became necessary to ensure continued and unrestricted access to the fishing grounds for the community, and to supervise equitable access and withdrawal rights to all fishers. It is also possible to see variations between different open-sea based villages as well as between different creek-based fishing villages. A comparison of the *Palle* fishing hamlet in Uppada with the *Vadabalija* of Uppada and the *Palle* of BCV Palem clearly shows adaptations that have many things in common with the *Vadabalija*.

The development of systems emphasising secure, sustainable and equitable access to resources

The assertion of rights of access and withdrawal, and their constant reiteration, provides a security of tenure to the fishers to the fishing grounds that can be equated in farming terms to acquisition of land tenure for cultivation purposes. Once that kind of ownership is established, it is to the advantage of the fishers to ensure that 'their' fishing grounds are protected not only from encroachments, but also in terms of their ability to deliver goods sustainably.

The issue uppermost in the Panchayat's role in fisheries management lay in ensuring equity of access. Ensuring equity is an important measure of management, because the confidence that the Caste Panchayat received from the community members depended on their belief in its just and equitable ways of allocation of access and withdrawal rights. It establishes and codifies a system of social norms and customs governing the sharing process, and plays a crucial role in conflict resolution in the process of sharing of the communal property.

Integrated and holistic nature of the systems of governance concerning resource allocation

By acting as a 'one-stop shop' to address most important needs of the fishers, the Caste Panchayat constantly reinforced its importance to them. Much of the fishers' faith and trust in the system comes from its delivery of a wide range of services beyond just fisheries. Although the purpose of this paper has been to highlight the fisheries management aspects of TCBMS, it is recognised that this function of the TCBMS must be related to their wider and more influential role in the community because, at the community-level, these functions are not divisible into separate categories, but form parts of an organic whole. Secondly, it is the way these various functions integrated that made the customary systems work, and they would need to be included in any study exploring the reasons for their viability and success. Obviously, any efforts to study specific features of the systems in isolation are bound to lead to misleading conclusions.

Participatory nature of decision-making

The Mofus Bandar monograph highlights the fishers need for a Caste Panchayat, because their 'problems have to be understood, their behaviour patterns and actions evaluated, rulings, directives and punitive measures administered by a social machinery which very closely participates in the community and is continuously subject to such rulings and thus know where the shoe pinches' (GOI, 1961). 'Participatory decision-making' is a key ingredient of most development and management programmes in the coastal areas, and so are ideas like 'empowerment', 'equity', etc., all of which were an integral part of several customary practices. One feature of the Panchayat system that stands out is the power with which the Panchayat could enforce a regulation in the community, all the time drawing its powers from it.

The detailed organisational structure in each village, and the differences each such system had with others in the neighbourhood indicate that the collectives began at the grassroots level, i.e., within a community, which gradually went on to form supra-organisations in order to provide legitimacy to, and recognition of, the fishing boundaries. This kind of 'bottom-up' organisation and functioning stands in contrast to the current fisheries management programmes, which originate from the top (often exceeding the national boundaries), and flow down as commands to the fishers at the ground level.

The administration of fishing systems such as shore-seines in Uppada and the estuarine nets in BCV Palem is highly decentralised, allowing members to take decisions in a mutually acceptable way. At the Panchayat level also, all problems and issues were discussed threadbare in the presence of all members. There was no time limit on the duration of the meeting, and discussions, which generally began early in the morning, could often go into the night. All members, irrespective of their interest in the particular cases being heard, were expected to attend, so their participation and acquiescence in arriving at the judgement at the end of the day was assured. Everyone had the right and the opportunity to discuss and debate all measures of management. All decisions are taken in the open – even if the final verdict is that of the *Peddalu* or *Pethandarlu* – and the respect for law on the part of the members stemmed both from recognition of its representative nature as well as from the awareness that it was the best of options available to the community as a whole.

Conclusion

Over time, the systems have come under increasing stress with the emergence of new, primarily – but not exclusively – state-driven, hence formally recognised, political, administrative and judicial systems. This stemmed from various factors: the post-Independence India, with its emphasis on achieving the socialist ideal through science and technology, disdained the customary systems (when their existence was recognised at all) as ineffectual, elitist and backward, hence took the path of imposing new – supposedly more egalitarian – structures to replace the old ones. Secondly, the increasing international demand for seafood opened up new opportunities in the fishing sector, hitherto considered a backward, subsistence-based, livelihood provider for the poor people, for introduction of new, capital-intensive, technologies, which were better able to exploit the resources. The new technologies inevitably changed the traditional systems within and without the communities. Thirdly, increasing links with the external world, influx of new players into fish trade, increasing population and industrialisation of coastal areas in general, have all had a role to play in undermining the traditional systems which were best suited to work in a close-knit environment with smaller groups of people. Together, these changes transformed the subsistence economy of fishing into a market-based one of international stature and significance, and the customary systems – fishing, fish processing & trade, and social & political organisations, including TCBMS – became a casualty in the process.

If the description of TCBMS as presented is rather uncritical about some of the assertions, it is certainly not because of an urge to idealise or glorify the past, but to understand the fundamental underpinnings of a way of life from the perspective of people who lived through it. This is not an attempt to justify the revival of the customary practices in their entirety, which is both impossible and impractical. Every epoch has its own ways of dealing with change, and from today's perspective, many of the customs and practices of the traditional systems left a lot to be desired.

The caste systems, for instance, were an efficient system to the extent that they codified and enforced a particular way of life, which evolved because of a particular set of factors prevailing at one time. In due course, as with any religion or system, the wood was lost for the trees, and caste and dogma took over from reason, and even long after the conditions that dictated a particular course of action became invalid, caste continued in a form that was both inequitable and constricting. They were efficient so long as the villages remained isolated or at least apart from one another, largely uninfluenced by the external world, but once this changed, the traditional systems lost their relevance.

Taking these limitations into cognisance, it has still to be accepted that there is much to be learned from the customary systems. The inequities that were a part of the system were also complemented by much that is admirable and, what is more, useful, and rejecting the systems in their entirety will be akin to throwing away the baby with the bathwater. In terms of fisheries management in particular, the traditional systems were built upon centuries of experience and understanding of the fisheries, and can certainly show a way forward. An understanding of the communities' management mechanisms and trying to build upon the customary systems, rather than imposing restrictions from the outside thereby inviting suspicion and hostility, would not only reduce tensions, but also makes for an effective implementation of the regulations.

References

Acheson, J.M. 1981: Anthropology of Fishing. Ann. Rev. Anthropol. 1981. 10:275-316.

Alexander, P. 1995: *Sri Lankan Fishermen: Rural Capitalism and Peasant Society*. Second Revised and Enlarged Edition, Sterling Publishers Pvt Ltd.

Bavinck, M. 1981: The Fortification of Petty Commodity Fish Production in North Sri Lanka: An Exploration. Doctoral Thesis, Free University: Amsterdam.

Bavinck, M. 1984: *Small Fry: The Economy of Petty Fishermen in Northern Sri Lanka*. Free University Press, Amsterdam.

Bavinck, M. 2001: *Marine Resource Management: Conflict and Regulation in the Fisheries of the Coromandel Coast*. Sage: New Delhi.

BOBP, 1980: Technical Trials of Beachcraft Prototypes in India, BOBP/WP/7 Bay of Bengal Programme: Madras.

Bowrey, T. 1993 (Reprint): A geographical account of countries round the Bay of Bengal, 1669-1679, AES: New Delhi.

Census of India, 1961: Village Survey Monographs: Monograph on Mofus Bandar, Srikakulam District, Directorate of Census Operations, Hyderabad.

Census of India, 1981: Survey Report on Village Mofus Bandar, Srikakulam District, Directorate of Census Operations, Hyderabad.

CSD, 1988: A report on techno-socio-economic survey of fishermen community in Andhra Pradesh. Council for Social Development for the Directorate of Fisheries, Government of Andhra Pradesh, 1988.

DOF, AP, 2000: Handbook on Fisheries Statistics of Andhra Pradesh, 1998-99, Government of Andhra Pradesh, Hyderabad.

DOF, Undated: Integrated development of traditional fishing in Northern Region of East Godavari district (Gadimoga – Pentakota): Project Report. Department of Fisheries, Andhra Pradesh, Hyderabad.

Francis, W. 1907 (Reprint 1992): *Vizagapatam District Gazetteer*. AES: New Delhi.

Government of India, 2001: *Handbook on Fisheries Statistics 2001*. Ministry of Agriculture, Fisheries Division, Government of India, New Delhi.

Hemingway, F.R. 1915 (Reprint 2000): *Madras District Gazetteers*: Godavari, Volumes I & II; Asian Educational Services: New Delhi.

Hornell, J. 1920: The origins and ethnological significance of Indian boat designs (read in abstract at the Lahore Meeting of the Indian Science Congress, January 1918), Memoirs of the Asiatic Society of Bengal, Vol. VII No. 3, pp 139-256, Calcutta.

ICM, 2000: Sustainable Coastal Livelihoods Project: Report on the Field Validation of the Literature Review Results: May-July 2000. Unpublished internal report of the DFID-funded Sustainable Coastal Livelihoods Project, implemented by IMM Ltd., UK.

Mackenzie, G. 1883 (Reprint 1990): A Manual of the Kistna District in the Presidency of Madras. AES: New Delhi.

Mathew, S. 1991: Study of territorial use rights in Small-scale Fisheries: Traditional Systems of Fisheries Management in Pulicat Lake, Tamil Nadu, India. FAO Fisheries Circular 839, FAO, 1991.

Paul Durrenberger, E. & Gisli Palsson, 1987: Ownership at sea: fishing territories and access to sea resources. *American Ethnologist*, Vol. 14 No.3, Aug 1987, pp. 508-522.

PHFP, 14. Introduction of an Improved fish smoking method in Andhra Pradesh, India. Chennai: DFID Post-Harvest Fisheries Project.

Rama Raju, B. 1978: *Folklore of Andhra Pradesh*. National Book Trust, India: New Delhi.

Salagrama, V. 1990: Where is Fish Smoked in India? Come to BCV Palem, Andhra Pradesh, in: *The Bay of Bengal News*, Volume I, Issue no. 39. Bay of Bengal Programme: Madras.

Salagrama, V. 2000: Small-scale Fisheries: Does it Exist Anymore? in: *The Bay of Bengal News*, Issue No. 16, April 2000.

Salagrama, V. 2002: Fish out of water: the story of globalisation, modernisation and the artisanal fisheries of India, in: *Proceedings of the Asian Fisherfolk Conference, 2002*, January 25-29, 2002, Prince of Songkhla University, Hat Yai, Thailand.

Schombucher, E, 1986: *Die Vadabalija in Andhra Pradesh und in Orissa; Aspekte der wirtschaftlichen und sozialen organisation einer maritimen gesellschaft*. Steiner Verlag Wiesbaden Gmbh, Stuttgart.

Srinivas, M.N. (Ed.) 1993: *India's Villages*, Media Promoters & Publishers Pvt Ltd., Bombay.

Srinivas, M.N. 1994: *Caste in Modern India and Other Essays*. Media Promoters & Publishers Pvt Ltd., Bombay.

Srinivas, M.N. 1995: *Social Change in Modern India*. Orient Longman, Hyderabad.

Stirrat, R.L. 1974: Fish to Market: Traders in Rural Sri Lanka. *South Asian Review* Vol 7, No. 3: pp. 189-207.

Stirrat, R.L. 1977: The Social Organisation of Fishing in a Sinhalese Village, *Ethnos*, 1977: 3-4, pp. 122-148.

Stirrat, R.L. 1988: *On the Beach: Fishermen, Fishwives and Fishtraders in Post-Colonial Lanka*. Hindustan Publishing Corporation: Delhi.

Suryanarayana, M 1977. Marine Fisherfolk of Northeast Coastal Andhra Pradesh, Anthropological Survey of India, Calcutta.

Thomson, K.T. 1989: Political Economy of Fishing: A study of an Indigenous Social System in Tamil Nadu. Thesis submitted for the award of the degree of Doctor of Philosophy, University of Madras, 1989.

Thurston, E. & K. Rangachari 1909: *Castes and Tribes of South India*. Asian Educational Services: New Delhi.

Tietze, U. (Ed.) 1985: *Artisanal Marine Fisherfolk of Orissa: Study of their technology, economic status, social organisation and cognitive patterns*. Vidyapuri, Cuttack.

Vivekanandan, V., C.M. Muralidharan, and M. Subba Rao, 1997: A study of Marine Fisheries of Andhra Pradesh, draft report.

Glossary of Telugu Terms

Agnikula Kshatriya (Pallè) Fishing caste in the central zone of coastal Andhra Pradesh (the fishing caste in BCV Palem village in this study)
Alivi Vala Shore seines (Uppada)
Dhoni Plank-built riverine and estuarine fishing boat, with a forward deck and tumble-home sides
Ethubadi Collections made by the Panchayat from individual members for funding community activities
Ethudu Kāpulu Estuarine dip net (BCV Palem)
Jāntādu Veta Drop lines (BCV Palem)
Jātaralu Village festivals, generally dedicated to the local deities
Kattadi Restriction or ban (imposed by the Panchayat)
Kula Kattadi Caste code *Masula* Stitched plank-built boat for open sea fishing and shore seining
Nāva Plank-built canoe in the central zone of coastal Andhra Pradesh
Padunu Peak fishing period in a fortnight (in estuarine fishing systems that depend on moon phases)
Pakkidevi Vala Estuarine drag net (BCV Palem)
Pallè (Agnikula Kshatriya) Fishing caste in the central zone of coastal Andhra Pradesh (the fishing caste in BCV Palem village in this study)
Pāllu (singular: Pādu) Shares (also membership to Caste Panchayat in BCV Palem) *Panchāyat* The council of five constituting the caste leadership
Peda Kāpulu (sing: Peda Kāpu) Literally, caretakers; designation of village elders in the Palle villages, also called Pethandārlu
Peddalu (singular: Pedda) Village elders (Panchayat leaders) in the Vadabalija villages
Peta Administrative unit of a Caste Panchayat in Vâdabalija communities
Pethandārlu (singular: Pethandāru) Literally, managers; designation of village elders (Panchayat leaders) in the Palle villages
Pillagadu Priest in Vādabalija villages
Pottu Vala Push net (BCV Palem)
Sabhyulu Members of the Caste Panchayat *Sambarālu* Village festivals
Sammita Village crier
Tappulu (singular: Tappu) Punishment meted out by the Caste Panchayat
Teppa Boat catamaran constructed by logs tied together
Ummadi Veta Community fishing – organised by the Panchayat to generate funds for community activities
Vādabalija Fishing caste in the northern and central zones of Andhra Pradesh (the fishing caste in Uppada village in this study)
Vala Net
Vala Kattu Stake net (BCV Palem)
Veli Ostracism

Map showing the study locations

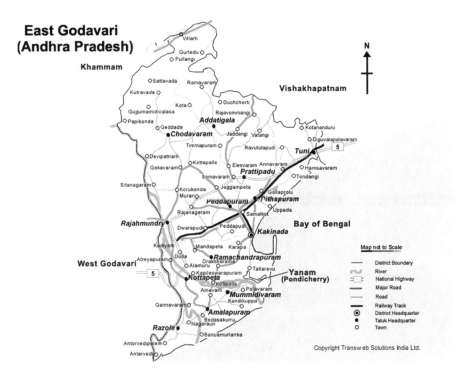

Source: BOBP Working Paper No. 46.

Self Regulation of Senegalese Artisanal Fisheries: a case study of Kayar

S.Y Alioune and J. Catanzano[1]

Introduction

Senegalese fisheries are currently characterised by the overexploitation of coastal demersal resources and the existence of excess fishing capacity. Catch per unit of effort is falling, with increasing numbers of fishing offences and increasing conflict at sea between fishers using different kinds of gear.

The seriousness of the crisis in the sector has been recognised by the Senegalese authorities, who since the beginning of 2000 have committed themselves to a fishery management policy based on the definition of mechanisms to allocate use rights. Fishery management plans will then be developed using these mechanisms in a manner appropriate to the chosen fisheries.

There is widespread recognition by all fishery stakeholders of the need to freeze fishing effort in order to implement mechanisms to regulate access to resources without discriminating by type of activity (artisanal or industrial).

The definition of use rights using different approaches (regulatory, co-management public auctions) with the best-suited tools (licences and/or quotas, other allocation methods) is currently a priority issue for the public authorities.

The legal and regulatory texts which currently underpin sector management include the possibility to develop and implement resource management measures and provide a basis for fisheries management, even if they remain inadequate and of insufficient breadth.

Law n° 98/32 comprising the marine fisheries code thus includes:
- The implementation of annual or multi-annual fishery management plans;
- The creation at national and local levels of consultative bodies comprising representatives of the administration and different socio-professional groupings.

However, in the absence of preliminary work to define access rights to the resource, these instruments cannot yet be put into practice.

It is in this unregulated context that localised fisher initiatives have emerged in

---

1 Respectively Fisheries Ministry, Dakar, Senegal; and IDDRA, Montpellier, France. The views expressed in this paper are those of the authors alone.

order to implement management measures in certain urgent situations: threatened collapse of the exploited fish resource, overproduction and dramatic price falls, and so on.

These spontaneous initiatives by some fishing communities illustrate the significant potential which exists to include communities but in a context where the legal framework is lacking, the fear is that such initiatives may be fragile with a risk of unsustainability.

The resource management system implemented in Kayar, the subject of this chapter, is one of such initiative.

Background concerning the village of Kayar

The village of Kayar is located on the northern Senegalese coast, 58 km from Dakar. The sea in this area is very productive in fish due to the favourable hydroclimatic conditions and the existence of a marine canyon.

The Kayar economy is essentially rural and depends on two main activities:
- Fishing, which is by far the most important activity;
- Market gardening, which is also well developed in the area.

The inhabitants of the village of Kayar are for the most part both fishers and farmers. Their income does not depend only on fishing.

The village population is estimated to be around 15,000 people.

The most recent census estimates there to be around 600 motorised pirogues operating from Kayar. Over the past 3 years, average landings are around 25,000 tonnes per annum, comprising sardinellas (during the season) and demersal species (breams, gilthead seabream, red pandora, bluespotted seabream, *thiof*, etc.)

Management system for artisanal fishing in Kayar

The management system for artisanal fishing implemented in Kayar developed in a context marked by the devaluation of the CFA franc. This devaluation impacted on the artisanal sector, particularly through a sharp increase in the price of inputs and fishing equipment, which are generally imported.

In Kayar, the 1994 devaluation did not however immediately lead to an increase in the landed price of exported fish (mainly red pandora). The price paid to fishers stagnated despite an increase in export prices.

Only those people involved in the marketing element of the fishing operation (traders, exporting factories) benefited from the increased export price as a result of the substantial increase in their operating margins. This producer price rigidity may be explained by the market conditions which existed on the beach in Kayar at the time, in particular:
- Many small producers faced with a small number of buyers of two types: independent traders and representatives of Dakar-based exporting factories. The latter purchased 80% of red pandora production and the market was dominated by one company (Amath Gueye).

- An excess of supply over demand. Each pirogue typically landed 10 to 25 15kg boxes of red pandora, sometimes leading to fish being discarded on the beach.

The landed price per 15kg box was set at 700 CFA francs (about €1.07 at 2003 rates) by the traders and did not change following the devaluation. The impact of the devaluation (with an increase in both input and general price levels) was to significantly worsen the revenue-cost ratio for the fishers.

The fishers reacted by calling a meeting under the village Head. The representative of the local fisheries service and village leaders attended the meeting. A local committee was set up and given the task of managing the situation.

Faced with a refusal by the fish traders to increase the price per box of red pandora, the fishers began a 3-day strike during which they stopped all fishing activity. When they re-commenced fishing, the fishers organised their own marketing, renting refrigerated lorries to transport their fish to the Dakar central fish market. A commission of designated fishers was established and was given responsibility for selling the fish. After each sale, the receipts were shared on a pro-rata basis between the fishers involved.

In the end, the determination and the perseverance of the fishers obliged the traders to begin negotiations with them on the price of a box of red pandora. The main buyer (Amath Gueye), who led negotiations for the traders, ended by accepting the fishers demand for a minimum price of 8,000 CFA francs (around €12.2 at 2003 rates) per 15kg box of red pandora, multiplying the price over eleven fold.

The deal that was negotiated also included the sorting of the fish so as to guarantee traders a marketable size.

However, this agreement, which was initially followed by the other traders, did not last long due to the persistent problem of an excess of supply over demand, which incited traders to break the agreement, paying lower prices to the fishers.

The local fisheries committee decided therefore to regulate production in order to support the minimum price. In order to do this, it fixed an individual production quota of 5 boxes per trip. This quota was later lowered to 3 boxes per trip in order to better match supply and demand.

This decision of the local committee was accompanied by a fine of 30,000 CFA francs (around €45.70 at 2003 rates) for over-quota landings. The non-payment of this fine led to the forced immobilisation of the pirogue of the fisher concerned.

The committee also adopted measures concerning the time and duration that the pirogues were allowed to go fishing. The aim was to ensure that the individual production quotas were respected and that the fish were sold at the agreed prices.

This set of measures has managed to keep the price of red pandora above the minimum of 8,000 CFA francs per box.

The organisational system which has allowed the effective application of these management measures relies on two commissions composed on fishers:
- A sales commission which has the monopoly on selling to traders;
- A control and surveillance commission which is responsible for the effective application of the agreement on all of the beach.

The actions of the local fisheries committee in managing red pandora landings have also had a positive impact on other fisheries. Purse-seine vessel owners, for instance, decided to manage their fishery which regularly suffered from overproduction due to intensive fishing effort. It was not rare to see pirogues undertaking three trips in a single day.

The impact of overproduction (unsold fish, drastic price falls) encouraged local fisheries to organise a structure "Mbalmi" to regulate trips. Fishing effort was restricted to one trip per day. The immediate effect of this measure has been beneficial as prices have risen substantially.

Factors leading to successful local management in Kayar

Implementation strategy
The strategy used by Kayar fishers comprised:
- Putting an end to the race for fish, which had led to substantial financial losses even as traders and factory owners were making substantial profits.
- Implementing an effective system to control the measures to regulate production which had been adopted.

In a Senegalese context characterised by a lack of organisation of production and marketing activities in artisanal fishing, this strategy was innovative in a number of ways. In particular, it led to:

Changes in marketing practice
The fishers tried a new approach which broke with the traditional methods. Previously, marketing of landed fish produce took place between the fisher and a trader with whom he was typically linked in one of a number of ways, most often: (i) through a family tie, (ii) through a monopoly given to the trader who funded the expenses of the fishing trip, (iii) through a long-standing working relationship with a particular trader.

In each case, this kind of marketing led to an asymmetric relationship between a multitude of sellers and a few buyers. It also led to opaque market arrangements and did not guarantee a reasonable price to fishers.

It is also necessary to recall the physical and mental state of a fisher at the end of a fishing trip, when he is not in the best shape to negotiate the sale of his produce. All these factors contributed to introduce price distortions.

The institution of a sales commission with a monopoly on the selling of fish represented both a novelty and progress in marketing practices in the artisanal fishery. With this new arrangement, the fisher was freed from the worries and problems related to selling his catch.

Progress in the organisation and management of artisanal fishing

The experience of Kayar fishers is the first example of structured and sustainable action undertaken by artisanal fishers to manage fishing locally in the sense of managing fishing effort.

It began a dynamic structuring process involving various fishing activities at the local level. New professional structures charged with regulating fishing emerged as a result. These include the "M'balmi I" and "M'balmi II" committees which concerning purse-seine vessel owner, the Kayar youth, and local sections of the national fisher and trader federations.

In February 2000, the local and national organisations linked to the fishing sector in Kayar formed an inter-professional Economic Interest Group called "yallay M'baneer ak feex gui".

The Kayar management model is held up as an example to other Senegalese fishers, as well as to fishers in neighbouring countries (Mauritania for instance). It demonstrates the possibilities for fisheries management by local communities. The important gains made in resource management in Kayar have led some people to propose the generalisation of this approach in Senegal and elsewhere.

Results obtained through the implementation of management measures

The implementation of management measures at the local level has led to the following results:
1. Increased landed value due to a substantial increase in landed price. Fisher incomes have increased.
2. Catch limits (individual catch quotas) and reduced fishing time have reduced pressure on the resource.
3. Improved quality of life for fishers due to the reduction in fishing time. This has also allowed Kayar fishers to devote more time to onshore activities.
4. Less intensive utilisation of fishing equipment (particularly motors) due to reduced fishing time. Working life of equipment thereby extended lowering direct fishing costs, and also indirect ones, resulting from the breakdowns associated with excessive utilisation.
5. Substantial savings in other production costs (fuel, boxes, ice).
6. Increased incomes have led to increased savings and an increasing capacity to fund fishing activities (fishers have been able to constitute reserves to fund the renewal of fishing equipment).
7. At the collective level, the dues paid to the local fisheries committee by the fishers have funded social actions: renewal of the dispensary, a fisher assistance fund.
8. Promotion of selective fishing in response to the sorting of fish imposed by the agreement.
9. Improvement in the beach environment due to the elimination of discards resulting from excess production.

In summary, the impact of the management measures has been positive at the biological, economic, financial and social levels as well as improving the quality of life of the fishers.

ANALYSIS OF EXPLANATORY FACTORS IN LOCAL MANAGEMENT SUCCESS

The particular features of fishing and of the environment in Kayar helped the implementation of "good" resource management practices. These features concern socioeconomic, ecological and human dimensions.

They relate in particular to:

The geographical features of the village of Kayar

From both marine and terrestrial perspectives, the village of Kayar is isolated. Road access is relatively difficult because it is located far from the main roads. Kayar is the only important fish landing centre in its coastal area.

This relative isolation of the village has two major consequences:
- The relatively sedentary nature of Kayar fishers who do not have the option of landing at other sites, as do fishers operating from other sites. Elsewhere, the decision by a fisher to land at a particular site depends on fish prices or on the numbers of traders present;
- Limited mobility of traders based in Kayar.

These two characteristics encourage understanding between different partners who almost have to organise and stabilise the market.

The human and social dimension

Kayar is an old village founded by people of the Lebou ethnic origin who are found along the Senegalese coast (especially the so-called "small coast" (la petite côte) and on the Cap-Vert peninsula). The Lebou have a reputation as well-established fishers and their society is based on a secular tradition of organisation based on strong and accepted social and political regulatory institutions.

Social cohesion and cultural homogeneity are the characteristic traits of this community. They encourage group solidarity based on actions which benefit the whole community. This explains the legitimacy of the local village fishery committee and the very broad adhesion to its objectives and actions.

The particular nature of fishing in Kayar

Fishing in Kayar is predominately line fishing, which is considered to be a clean and selective method. Other fishing gear (drift nets, monofilament nets, purse seines) are in the main used by fishers who are foreign to the village (from Saint Louis), who are either established in Kayar or who undertake seasonal migrations.

In defined marine zones, the community has forbidden the use of certain gear (drift nets). These bans lead to recurrent conflicts between local and migratory fishers. At the same time, they demonstrate the use of traditional management methods in Kayar and the will to implement improved management of both landings and fishing methods.

The proximity of the fishing grounds is another feature of Kayar. It provides the following benefits:
– Lower fishing costs (fuel savings);
– Opportunity to land fresh fish with a "catch of the day" quality label for which there is strong international demand (especially from the European market).

Landings of red pandora are particularly important in the region. They represent on average some 20% of total national catch (annex 3). The favourable ecological conditions explain the abundance of demersal species (thiof, sole and the like).

Taken together these factors explain the attraction of Kayar to traders and in principle put fishers in a strong position in negotiations with traders.

The dominance of line fishing and short daily trips also characterises fishing in Kayar. These aspects are important taking into account the difficulty of regulate catch and check landings for other fishing types, for instance, in the case of line fishers using ice. In the case of this latter, the main difficulty concerns the large number of possible landing points. For this type of fishing, the landing point is not determined in advance, and the strategy of the fisher depends on a number of factors, particularly:
– The price of fish on the beach;
– Their reserves of ice and fuel;
– The number of traders on different beaches etc...

This adaptability on the part of fishers is a management constraint because of the difficulties caused in monitoring and control. In Kayar, this constraint is absent.

The role of the fisheries administration and the public authorities
Although an initiative of the fishers, the management system has benefited from the support of the local administrative services.

The understanding and indeed the active participation of the fisheries administration in the system has been a key factor leading to success.

In the Senegalese context, fishing regulations do not, in principle, allow local communities to take measures to restrict fishers' catches. The institutions in charge of fisheries regulation and administration thus have no legal foundation to accept or support this kind of experiment given the current fisheries law.

At the beginning of the strike movement by fishers, legal proceedings were initiated against the leaders of the local fisheries committee who were accused of preventing people from working. It was only the relatively minor nature of the dissidence that prevented the Kayar fishers' action from leading to a major judicial conflict with unforeseeable consequences.

The "illegal" nature of the Kayar management measures limits their generalisation to the rest of the country. The measures represent a potential source of conflict if they are applied to areas not having the same characteristics as Kayar.

The role played by the local fisheries administration was also decisive in the area of assistance and advice.

Despite their expertise in fishing, most Senegalese fishers are illiterate. This represents an obvious handicap when it comes to planning a widespread action and to resolving problems in the battle with traders, who are a more structured group with stronger financial resources.

It is for this reason that the assistance and advice provided by the local fisheries service was beneficial. This assistance concerned in particular legal advice to help prevent fishers from seriously infringing the law.

Assistance also came in the form of propositions to resolve the difficulties met for instance in hiring refrigerated vehicles to transport the production to the central fish market in Dakar.

Moreover, the action undertaken by the fishers went beyond the strictly local framework and involved structures representing the State. Problems were initially raised with village-level institutions. The first meeting was called by the village Head who administratively represents the State and all populations in the locality.

The involvement and support of this local authority legitimised the Kayar fishers movement and gave them a stamp of legality.

External and exogenous factors in the implementation of local management

The case study has shown the importance of external and exogenous factors in the action of local fishers.

External events were the major factor initiating the implementation of management measures. The currency devaluation followed by the failure to adjust producer prices had a profound impact on the financial equilibrium of fishers, leading to large-scale rent capture in only one element of the fishery sector.

Hence, management measures were a reaction to an exogenous shock. They were not the result of a proactive process seeking to implement sustainable resource management. This analysis raises the question of the limits to the Kayar management system.

- The influence of external factors is also due to the openness of Kayar fishers to the outside world through the relationship that they have long had with other fishers at the international level, particularly European. Their network enabled them to make frequent visits abroad and to learn about international attempts to manage the fisheries sector.
- At the beginning of the 1990s, Kayar fishers visited le Guilvinec in France to learn about the fish marketing systems which are used there. They often mention the importance of such trips in the development of their vision for the collective organisation and management of their professional activity. Whilst not determinant, this factor certainly played an important role in the initiative undertaken by the fishers.

CRITICAL ANALYSIS OF THE RESULTS OF LOCAL MANAGEMENT

Graphs 1 and 2 (annexes 1 and 2) depict respectively the evolution of annual red pandora production and the number of line pirogues in Kayar. These graphs show:
- A downward trend in catch of red pandora over the period 1999 - 2002. Between 1994 (first year of the management measures) and 1995, the fall in production was 20%. And over the period 1994 - 1999, the fall was around 27%;
- A substantial increase (30%) in the number of pirogues between 1994 and 1995. In 1996, the number dropped back but remained high compared to the 1994 reference year. Looking at both pirogues and landings, it is clear that the former increased whilst production began to fall. In 1997, the number of pirogues fell. One explanation put forward concerns the reduced attraction of Kayar for migrant fishers who preferred to use other landing points. The reduced catch of red pandora is the main cause. Both 1997 and 1998 saw large increases in the number of pirogues due to the exceptional abundance of octopus in Kayar. Since 2000, things have returned to normal.

These developments may be interpreted as showing:
1. A first period (1996-1999) during which the price effect led to an increase in fishing capacity in terms of number of pirogues, followed by
2. A period of contraction in the number of pirogues as they adjusted to reduced catch, and
3. A period of (sustainable?) reduction in red pandora catch over the past four years, and which could be interpreted as a sign of resource overexploitation. This reduction has occurred despite the decrease in the number of pirogues.

These results show the limits of the catch control system put into place following the de-regulation of the red pandora market. The study of this system shows that the mechanisms used to determine quotas do not relate to an overall TAC, and do not represent an element of resource rent capitalisation within an access control framework.

The level of individual catch has been determined empirically using a trial and error method which has not been costless since the first attempt (which set the quota at 5 boxes) led to overproduction. The aim of the scheme is to set supply to the level that maximises producer prices.

Implicitly this kind of management seeks to manage output so that increases in demand do not lead to overproduction and hence future declines in prices. The gains are capitalised by controlling price increases and are shared equitably through the quota system.

But even though the individual quota has remained constant, the increase in the number of pirogues, attracted by the relatively high profit levels, has created an overexploitation dynamic in the fishery. This problem shows the need to control access to support the quota system.

In the absence of access control, Kayar has suffered from this problem. The price of a box of red pandora has sometimes reached 13,000 CFA francs (around € 19.80) and the number of pirogues grew appreciably. Red pandora catch on the other hand has fallen over the past four years.

These results show that in the absence of access control, the approach followed by the Kayar fisheries committee does not ensure sustainable resource management. In the short term, the implementation of individual catch quotas had a positive effect on the resource due to the decrease in total catch. But in the longer run, the dynamics of fishing capacity have led inexorably to increased pressure on the resource.

Limits to the Kayar artisanal fishery management model

Limits related to the management objectives

In practice, the goal of managing to achieve sustainable resource exploitation was not the primary goal of the Kayar fisheries committee. First and foremost, the aim was to respond to the fall in real incomes suffered by the fishers after the devaluation of the CFA franc.

At the same time, through the implementation of individual catch quotas, the initiators of the movement thought (or suffered from the illusion) that they were sustainably managing the resource.

Current results from fishing in Kayar show, as everywhere else in Senegal, increasing signs of the overexploitation of coastal demersal resources. Average catch rates per pirogue have fallen substantially since 1994.

During discussions with senior members of the Kayar fisheries committee, they expressed the view that the system to organise and control of landings has seriously declined over the past few years. The sales and monitoring commissions instituted by the movement are virtually inactive.

When asked as to the causes of this decline, Kayar fishers reply that the abundance of red pandora observed at the time have not existed for a while and that it is increasingly rare to see a fisher catch his 3-box quota.

The usefulness of these institutions is no longer perceived. Local fishery management measures conceived for a single species and without controlling entry into the fishery have served their time and are no longer sufficient to ensure sustainable resource management.

The downward trend of total production and individual catch rates, indicative of resource depletion, show that the Kayar fishery management system was not able to ensure long term sustainable resource management. The limits related to management objectives relate to the failure to take into account the dynamics of fishing capacity.

As everywhere in the case of artisanal fishing, Kayar fishers resist measures to limit the number of pirogues or fishers, because such measures appear to conflict with their view that fishing should be an activity open to one and all.

They feel that catch limits should apply exclusively to vessel owners who do not come from the traditional fishing community (for instance, civil servants or traders who have invested in the sector). No limit should apply to fishing crew, whether they come from Kayar or not.

The monotonic increase in fishing capacity in Kayar fits with this vision. The increase arises from a number of forces:
- The demographic growth of the village, which has reached high rates. It is the tradition in the village that sons of fishers automatically become fishers;
- The natural tendency for fishing crew to buy their own vessels, either because they are becoming independent of the family exploitation, or because they have saved enough to create their own activity.
- The tendency for fishers to increase their fleet by buying new vessels as their incomes increase.
- The absence of alternative employment in Kayar (apart from market gardening) which makes fishing the employer of last resort;
- The existence of a large community of migrant fishers who are either seasonal or permanently established in the village.

This recurrent increase in fishing capacity increases pressure on the resource and undermines the local management efforts made by fishers. They feel that they are sustainably managing the resource because they do not recognise the limited impact of individual quotas if fishing capacity is not limited.

Limits related to institutional and structural weaknesses of local management
The limits of the Kayar experience are also found in the absence of institutions which must necessarily accompany regulation. The absence of research has been a particular problem in a number of areas discussed in the remainder of this section.

Prior to the implementation of the regulation given that individual quotas have been determined empirically on a trial and error basis. To begin with, quotas were set at 5 boxes but this had to be reduced to 3. Had research been involved, it would have been possible to calculate individual quotas with reference to a TAC which would have offset the lack of access control. This approach, set within a context of sustainable resource management, would have been different to the Kayar approach which fixed quotas with the sole aim of maximising producer prices.

But the implication of research in the calculation of sustainable yields only makes sense if it is conducted first at a national level. Otherwise, applying research results at a purely local level raises other problems. Nonetheless an empirical approach could have been adopted to associate research with the setting of an overall TAC for Kayar, based on an assessment of available biomass. This example shows that research remains ill-adapted to accompany local initiatives, which are inevitably partial at least to begin with. But undertaking a joint action with research would have allowed for training and for the gradual evolution of the initiative towards the idea of controlling capacity together with setting individual quotas.

By way of example, the high price for red pandora in Kayar led to an increase in migrant fishers increasing the number of pirogues by 30% between 1994 and 1995.

In the monitoring of management measures, the involvement of research would have allowed the undertaking of scientific resource assessments. It would also have enlightened fishers as to the biological and economic interactions that could undermine local resource management in a context of uncontrolled capacity.

The lack of support from the national MCS institutions was also a handicap. Fishers generally managed to control land-based activities (landings and prices) but they could not be certain that infractions were not being committed at sea by either artisanal or industrial fishers.

The main lesson is that local management measures were not integrated into an overall management system that defined in particular MCS arrangements on land and at sea. The community based management institution and the national MCS institution did not plan their activities together, so the lack of synergy was not a surprise.

POSSIBILITIES FOR THE GENERALISATION OF THE KAYAR CASE

In the post-devaluation period, Kayar was not the only place faced with a dramatic fall in real producer prices. Fishers at other sites were also hit by this problem (for example at Mbour, Joal, Soumédioune etc.). But the same reaction did not occur, at least not in the long run.

In Senegal, there are occasionally initiatives in the artisanal fishery sector to regulate fishing effort or production. Such initiatives include limits on trips by purse seiners at Mbour, Joal or Saint-Louis.

The octopus season regularly features incidents between traders and fishers over price formation, leading the fishers to restrict their own catch.

But these restrictions are not sustainable for a number of reasons, in particular:
- Inability to apply catch limits comprehensively due to control problems (due to the many landing points);
- the tendency for fishers to give up the management measures once prices return to their normal level.

The particular feature of the Kayar case is that it is based on an organised and collective movement on the part of the fishers that has proven sustainable. The various features of Kayar noted above explain this result.

The geographical and human profile of Kayar is found in other fishing areas in Senegal, for instance, villages situated on the Saloum Islands which are likewise isolated.

In these villages, artisanal fishers specialise in shrimp fishing. In Bettenty, the fishers have, with the support of the fisheries administration, implemented shrimp fishing regulations and created an organisation to ensure compliance (creation of a beach committee responsible for enforcing the adopted rules).

In this case the approach is to control mesh sizes, and to ensure compliance with the opening and closing dates of fishing seasons. This action was taken because of the increasing numbers of juvenile shrimp in the catch. The community, recognising the importance multiplier effects that the shrimp fishery has on the economic and social life of the village, acted when faced with the threat of resource depletion.

In this case, MCS has been effective and has produced beneficial results. The size of catches has increased and the selling of the catch takes place in conditions which are much more advantageous for the fishers.

In the local context, regulation of the artisanal shrimp fishery should exist in any case but it is not applied due to the inability of the fisheries administration to ensure compliance. This failure may be explained by the prohibitive nature of enforcement costs and also by social and political pressures which tend to paralyse action by the administrative services.

It is the action of the fishing communities themselves that have allowed the situation to be changed radically. A parallel with the Kayar situation exists in the following aspects:
- The realisation on the part of the fishers of the threat they faced and their reaction to it;
- The importance and effectiveness of the institutions established and managed by the fishers themselves (beach committee in Bettenty, local fisheries committee in Kayar);
- The greater effectiveness of community institutions over State ones in the management of fish resources.

Even if the management objectives differ, the two examples show the possibilities for success in resource management. But it is important to note that implementation conditions were similar in the two cases.

Useful lessons for Senegalese fisheries management

Capitalise on succesful experience in fisheries management

The Kayar experience contains elements that are useful to help progress current discussions and work on fisheries management in Senegal. The realisation by fishers of their ability to influence certain key variable (price, catch) through an appropriate organisation and management system is a favourable element to increase community responsibility in artisanal fishery management.

In other contexts, this may seem an unremarkable advance. But in Senegal, it represents a qualitative change of great importance in the development of the artisanal fishery and the mentality of the fishers.

The nature of industry institutions responsible for managing fishing rights

In Kayar, the fundamental criterion for the organisation of fishers is the community involved in fishing practices related to a specific fishing gear. The red pandora committee responsible for red pandora catch is made up exclusively of owners of line-fishing pirogues.

Similarly, purse-seine owners are organised into their own structures (Mbami 1 and Mbalmi 2) for Kayar pelagic fishing.

This preference of fishers for specialised institutions expresses the differentiation that the fishers perceive amongst themselves and gives them an identity.

In diagnosing the ills facing the fishery, artisanal fishers often point to the role played by certain unselective fishing gears (purse and beach seines) or the poor exploitation patterns that they encourage (set nets, drift nets, monofilament etc).

The expression of these problems causes vivid opposition between fishers and mutual accusations of responsibility for fish resource overexploitation.

These gear-based structures provide perhaps a way forward for the development of industry organisations who will be responsible for resource management. In any event, they provide guidance to future study of the question.

A fishery-based management approach

In Kayar, the application, monitoring and control of management measures became less complicated and easier to implement once they applied only to one fishery. The propensity for fishers to be able to pose and resolve resource management problems at the scale of a fishery should be taken into account in future discussions on appropriate institutional arrangements.

Defining management space

In the case studied, the spatial level of regulation from an administrative viewpoint is the village. A year ago, this village became recognised as a commune. The definition of a relevant terrestrial management space is important. For the exploited species (red pandora), it could be considered that the marine space is a continuation of the terrestrial space. This is explained by the proximity of the fishing zone and by the isolated nature of Kayar compared to other fishing sites.

In the case of other resources, such as the pelagics, the marine area concerned is vast and covers numerous landing points.

In a place such as MBour (one of the largest fishing centres in Sénégal) there are common fishing grounds between line-fishers from MBour and from other nearby villages. In such a context, it would be illusory to envisage resource management based on a single community. No single locality can claim to have exclusive management jurisdiction over a particular marine space.

Apart therefore from exceptional cases like Kayar, a management system aiming at price regulation would only make sense at the national level, or at the very least by combining a number of regions into a single system.

Area of competence of the management system

The area of competence of community management institutions proved very limited in the case of Kayar. These institutions were not concerned with access to the resource. But the goal of sustainable management cannot be achieved if the problem of free and open access in artisanal fisheries is not addressed.

By planning trips for line-fishing or purse-seining pirogues, Kayar fishers manage to control fishing effort. This effort control is considered by them to be an effective means of resource management. But because fishing capacity remains uncontrolled, the local fisheries committee cannot ensure resource conservation. The paradoxical situation has then arisen that resource management measures have led to a fall in landings of red pandora in Kayar.

Accompanying measures necessary for local management
The analysis of the limits to the Kayar model shows the needs for information, training and awareness-building amongst artisanal fishers concerning fisheries management and the significance and limits of different management tools.

At the same time, these needs reveal the limits to the advice and assistance given to Kayar fishers by the local fisheries administrative services. Such assistance has not been guided by the principles of fisheries management. It is characterised rather by reaction to movements initiated by the fishers.

Overall, this experience shows the amount still to be done to train people (in the administrative services and NGOs) who can provide support to community-based management institutions.

Dealing appropriately with migrant fishers
The problem posed by migrant fishers in the implementation of the management system was resolve harmoniously. The fisheries committee was open to such fishers who were fully integrated into the movement. In the end, the problem of migrant fishers did not prevent the application of fishery management measures at the local level.

Internalisation of regulatory institutions
One possible explication for the failure of attempts to regulate production in large landing centres such as MBour is the absence of legitimacy on the part of the village head to be the reference point, to settle disputes and to build consensus on a course of action.

In the absence of such an authority, it is the administration which is the last resort. But it appears that the administration is ineffective due to the lack of policies and management instruments to which it can refer.

The existence of traditional or modern, decentralised institutions with which the communities can identify is an important element in the success of community management systems.

Definition of a framework and conditions for local management
Resource management raises the question of access, a question which has to be settled nationally before it can be settled locally. In order to do this, it is necessary to define fishing rights and their method of allocation together with the arrangements for their monitoring and control.

Senegal has been undertaking this task for over a year and is close to the end.

The process is based on an urgent need and the will to freeze fishing effort so as to facilitate the implementation of resource regulation without discrimination between artisanal and industrial fishers.

The mode of regulation must have transparency and equity as guiding principles, must be in accordance with development goals and must be coherent with national and international law.

These considerations lead to the conclusion of the need to integrate local experience into nationally-defined management policy. The only way to achieve this is:
- Local management based on nationally-defined management principles;
- The co-ordinated and coherent intervention of institutions that support fisheries management (research, MCS, administration). This will also help to provide an improved framework for the operation of NGOs and fisher support services (fisheries service).

Giving responsibilities to fisher communities

One lesson from the Kayar experience is that fisheries management has more chance to be accepted and to succeed if it is based on the will and decisions of fishers themselves.

For a long time, numerous researchers have maintained the impossibility of implementing individual quotas because of the difficulty that the administration would have to enforce them.

The example of Kayar (even if it is difficult to generalise) shows the possibility to use a fishers' collective organisation as a way of implementing such measures. If it can be done, enforcement of measures by fishers is cost effective.

The multi-specific nature of the fishery did not prevent the introduction of catch-based measures. The case of Kayar shows how the market dominance of one species (red pandora) can lead it to play a determinant role in the fishery's dynamics. The whole regulatory system developed around this one species with relatively positive results.

References

« Galgui », revue d'informations sur la pêche artisanale maritime au Sénégal du Collectif national des pêcheurs du Sénégal (CNPS), numéros 3 (août 94), 4 (décembre 94).

FENAGIE Pêche: Gestion rationnelle et durable des ressources halieutiques du Sénégal: cas du comité de pêche de Kayar – Document de la fédération nationale des GIE de pêche du Sénégal.

Ministère de Pêches. 2001: Stratégie de développement durable de la pêche et de l'aquaculture.

Catanzano. Joseph and Aly Samb. 2000: Réflexions pour une stratégie opérationnelle et programme d'actions prioritaires pour le secteur des pêches maritimes, Sept. 2000, réf. TCP/SEN/8925.

Annex 1

Table 1: Annual red pandora landings in Kayar and relative importance of the site in artisanal Senegalese landings

Year	Quantity (tonnes)	% of total landed by Senegalese artisanal fishery
1 994	2 309	29%
1 995	1 832	27%
1 996	1 450	12%
1 997	479	11%
1 998	749	21%
1 999	1 693	16%

Source: CRODT census

Graph 1

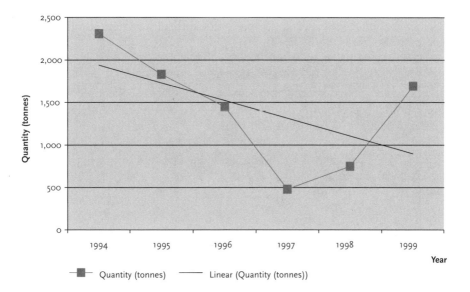

Annex 2

Annual number of line-fishing pirogues (reference month: February)

Year	Number
1994	520
1995	690
1996	610
1997	398
1998	731
1999	623
2000	482
2001	490
2002	458

Source: DPM

Graph 2

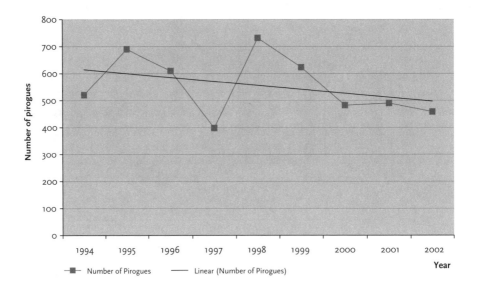

Annex 3

Monthly and annual variations in price of a box of red pandora

Year/Month	January	February	March	April	November	December
1993	1 815		5 070	3 855	2 265	2 430
1994	4 380	5 400	7 770	7 215	3 990	4 830
1997	8 190	6 450	7 200	9 030		
1998	7 260	12 375	13 935	7 125	7 500	5 985
2000	6 030	6 300	9 870	9 420	3 390	5 400

(Source CRODT)

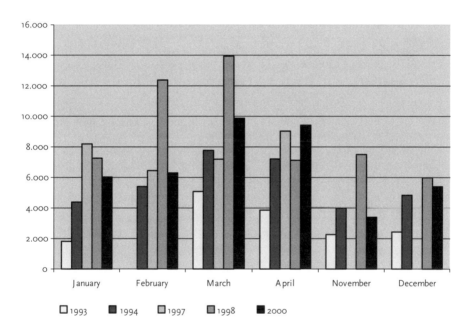

The Namibian Hake Fishery
Peter Manning[1]

Introduction

Namibia, a former German colony, was administered from 1920 by South Africa under a League of Nations mandate. The UN General Assembly in 1966 found that South Africa had not fulfilled the terms of the mandate and that the mandate had been terminated (UNGA 1966). This position was endorsed by the UN Security Council in 1969 (UNSC 1969), and Namibia was declared a 'Trust Territory' of the UN. The legality of the Security Council's action was confirmed by the International Court of Justice in 1971 (ICJ 1971). However, it was not until 1990 that Namibia became independent.

The pre-independence management of Namibia's fisheries was profoundly influenced by the complex array of political-legal issues as a result of which two distinct regimes emerged. The inshore fishery, over which South Africa exercised some measure of control, arose from South Africa's de facto jurisdiction over Namibia. The offshore fishery, over which neither South Africa, as the de facto authority for Namibia, nor the United Nations as the de jure authority, were able to exercise jurisdiction, was in effect an open access fishery.

The illegal nature of South Africa's presence in Namibia prior to 1990 meant that, when the South African Government proclaimed a 200 nm EEZ for Namibia in 1981, the zone did not receive international recognition and the South African Government was unable to enforce it (O'Linn and Twohig 1992).

The significance for the Namibian fisheries sector of Independence was that the new state could effectively assert jurisdiction over the rich off-shore fishing grounds that had hitherto witnessed an international free-for-all. This included the valuable hake fishing grounds.

The nature of the coastal environment - the Benguela ecosystem

The highly productive Benguela marine ecosystem, on which the Namibian fisheries are based, is one of the four major eastern ocean boundary upwelling systems of the world. The oceanography of the south-east Atlantic region is similar to that of the Humboldt current off Peru and Chile in many respects. The cold Benguela Current, a relatively narrow jet current of largely upwelled water flowing northwards at 1-5 km per hour (Field and Glazewski 1992), influences the waters above the continental shelf from the west coast of South Africa in the south to just north of the Kunene River, which forms part of the Namibian border with Angola.

---

1   Peter Manning is an independent fisheries consultant.

The barren, hyper-arid Namib Desert stretches the full length of the 1600 km long Namibian coast and for some 100 to 150 km inland. The coast has few bays or natural harbours and, in the central region south of Walvis Bay where the Namib sand-sea is found, large sand dunes sweep right down to the shoreline, making road or rail construction impossible.

This inhospitable barrier between habitable land and the sea has meant that very few of Namibia's people historically lived on the coast. Lack of an historical association with the sea in turn has meant that, despite the richness of Namibia's marine fisheries resources, the Namibian people have no tradition of exploiting or consuming their marine resources and of claiming a long standing traditional right to the resource. Thus it is that Namibia does not have an established artisanal fishery, so typical of most developing country fisheries.

The combination of desert and the typography of the coastline has meant that Namibia has only two harbours, those of Walvis Bay and Lüderitz, and no other significant landing sites, thus limiting the places where fish can be landed and increasing the ease with which the fishery can be controlled.

Environmental variability

Namibia's marine environment is characterised by periodic major environmental perturbations. Managing a fishery through such environmental events becomes critical to the survival of many fisheries and thus the degree of responsiveness to such events is integral to successful management.

In Namibian waters, warm water intrusions, which are part of the El Niño/Southern Oscillation (ENSO) phenomenon, happen periodically, as does the better known ENSO warm event in the Pacific. Because of their similarity to El Niño events in the Pacific Ocean, Shannon et al. (1986) termed these events the Benguela Niños. Major events were recorded in the Benguela system in 1934, 1949, 1963, 1984 (described by Shannon et al. 1986) and 1993-5 (O'Toole 1995).

When the warm event is peaking in the Indian and Pacific Oceans, the south to south easterly winds productively pump the upwelling system off Namibia. The south easterly winds weaken about 12-18 months after the Indo-Pacific El Niño occurs, the upwelling process weakens, sea surface temperatures rise and the Benguela Niño is underway (Anon. 1994, p585). A prolonged warm water event occurred in the southeast Atlantic from the latter part of 1993 until 1995 following the extended two year El Niño (1991-1993) in the Pacific; this had a considerable impact on the fisheries.

These events could be seen as an ocean equivalent of a terrestrial drought, where the plankton dies off and there is mass mortality up the food chain. Such an event interrupted the steady improvement in Namibia's fish stocks between 1993-1995 where the apparent mortality rate of the hake stocks was not explained by fishing mortality alone.

Managing the fisheries through such events is a critical part of the challenge of good fisheries management. This necessitates not only recognising the scientific message signalling the need drastically to cut the catch, but also having the economic flexibility in the industry to do so. Optimal fleet and processing capacity, from an economic perspective, may be less than what is required for peak years, depending on prices for product and the costs of harvesting and processing (Manning 2001).

Namibia's principal fisheries

In 1990, the then newly independent Namibia inherited a fisheries sector in which the most valuable commercial species had been over-fished and were in a depleted state.

About 90% by mass of the total catch of commercially exploited species fall into three major resource groups. Epipelagic shoaling species, pilchard and anchovy, are found inshore and are harvested by the purse seine fleet. The semipelagic Cape horse mackerel are harvested mainly by mid-water trawlers (a portion of the catch are harvested as juveniles in the purse seine fishery), and the hakes are the main species taken in the demersal fishery. The smaller monk and sole trawl fishery has a hake bycatch. A relatively small deepwater trawl fishery harvests mainly orange roughy. The most important crustacean fisheries are those exploiting the deep sea red crab and the Cape rock lobster (Bianchi 1999).

The total catch of all species since independence has varied between about 500000 to 800000 tonnes per annum. The catch reached a low point in the mid-1990s during a particularly severe Benguela Nino event. The total catch of all species was 623786 tonnes in 2000.

The fisheries and the Namibian economy

The contribution of the fisheries sector to GDP rose from about 4% at independence to 10.1% in 1998, despite the environmentally induced downturn during the mid-1990s (NEPRU 2002). This contribution to GDP was made up of about 4% from the harvesting sub-sector and 6% from fish processing (ibid).

About 95% of Namibia's total fish production is exported and the value of these exports in 1999 was about N$2.3 billion (US$292.621 million[2]) (ibid.). Fish and fish products contributed about 30% to total export earnings, with the demersal species bringing in about 84% of total earnings from fisheries (ibid.).

About 14220 people are employed in the fisheries sector in Namibia, approximately half of whom are employed in onshore processing.

The hake fishery

The hakes are by far the most important of Namibia's commercial fisheries resources. Three species of hake are found in Namibian waters one of which, *merluccius polli* or Benguela hake, is confined to the far northern areas of the Namibian EEZ and makes up an insignificant portion of the catch.

---

2 June 2003 exchange rates.

The hake fishery is focused on the two other species: cape hake, *merluccius capenis*, and deep water hake, *merluccius paradoxus*. Both species are found from close inshore to depths of 900m off the coast of southern Africa, although Cape hake are more commonly found between 100-350m and deep water hake are found principally between 300-500m. The young of both species are found inshore (between 25m -100m isobaths) during their first year before migrating to deeper water (Punt, Butterworth and Martin, 1995).

A BRIEF HISTORY OF THE HAKE FISHERY

The pre-Independence period
Some understanding of the history of the fishery is necessary in order to appreciate how the present fisheries regime developed, and the extent of the success in seizing the opportunity created at independence of taking control of a major natural renewable resource.

The Namibian hake fisheries began to rapidly developed with the arrival of foreign trawlers from Spain and the USSR in 1964. During subsequent years vessels from many distant water fishing nations entered the Namibian hake fisheries. The Namibian fishing grounds were regarded as international waters because of the illegal nature of South Africa's occupation of Namibia and the consequent non-recognition of the EEZ declared by South Africa for Namibia.

The hake catch grew rapidly from the 47 600 tonnes in 1964, to the 815 000 tonnes declared in 1972, the highest hake catch ever declared for Namibian waters (ICSEAF 1983). There was a general downward trend in catches after that until 1980, when the declared catch was 156 300 tonnes (ibid.). Catches then again rose until 1985 but declined throughout the rest of the decade.

The International Commission for South East Atlantic Fisheries (ICSEAF) was established in 1969 in an attempt to introduce some management of the fisheries. With the declaration of EEZ by neighbouring states in the 1970s, the primary interest of ICSEAF was the waters now under the national jurisdiction of Namibia. When Namibia became independent in March 1990, there was little purpose in ICSEAF continuing and the decision was made on 19th July 1990 to terminate that Convention[3]. ICSEAF's was not able to establish any effective system of monitoring the catch, let alone of surveillance and control. At the time of independence the hake stock had been depleted (Bianchi et al, 1999).

The Post-Independence period
Reducing the catch
The hake fishery, which had been overwhelmingly dominated by foreign fleets,

---

3 ICSEAF has not yet been formally ended as the Protocol of Termination only comes into effect when all contracting parties have deposited instruments of acceptance of the Protocol. Thirteen countries have not yet done so. See: http://www.fao.org/Legal/treaties/madrid2e.htm

experienced dramatic changes in the post-independence period. The new Government of an independent Namibia put an end to the high rate of exploitation of the hake stocks of the pre-independence period. Total annual reported landings of both species had declined from about 300,000mt in 1987–89 to about 130,000mt in 1990. The new Government then decided to cut the TAC to 60,000mt in 1991 to encourage a process of stock recovery (see Figure 1).

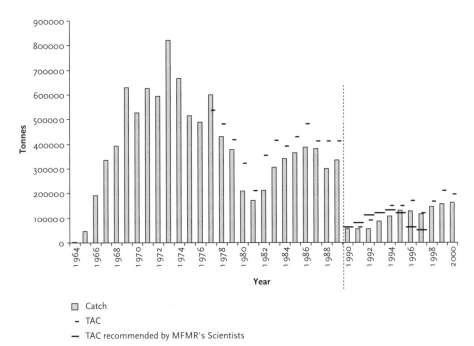

Note: showing TAC recommended by scientists, where available, and actual TACs. The dotted line indicates the date of independence, 21 March 1990.

Note that from 1965 to 1989, according to ICSEAF statistics[4], 10 664 600 tonnes of hake was reported to have been removed from Namibian waters by foreign fleets, valued at 2001 prices at US$28219 million[5]. Namibia received no share of this wealth.

Asserting control
The new Government declared an exclusive economic zone (EEZ) in June 1990, and ordered unlicensed foreign vessels to leave the fishing grounds. When a signif-

---

4   ICSEAF statistical Bulletins 1983, 1986, 1987, 1989. Note that ICSEAF Divisions 1.3 and 1.5 overlap into Angolan and South African waters respectively but by far the bulk of the catch can be assumed to have been caught with Namibian waters where the stocks are mainly located.
5   Average ex-coldstore price in 2001 for the seven size categories of Namibian hake in Spain from Globefish European fish price reports.

icant part of the Spanish fleet defied the order, the Government took tough enforcement action, despite being perceived as having very little capacity to do so at the time.

This action is important to note as it provided an immediate signal that the Government was serious about enforcement, i.e. it displayed a capacity for "strong government". It did so aware that key European Commission officials at the time strongly opposed such action, but also aware that the health of the fish stocks was more important for Namibia's economic prospects than European aid disbursements[6].

It also demonstrated that fishing companies, under certain circumstances were ready to seek, and even finance, enforcement by Government. The new fisheries administration in Namibia[7] at independence was left with one aging patrol vessel and no aircraft to patrol 500 000 sq km of water. After initially heeding the call to leave Namibian waters, about 40 Spanish freezer trawlers returned to the fishing grounds, believing the Namibian Government did not have the enforcement capacity to do anything about illegal fishing. The Government, realizing that, if they did not take action, more of the vessels that had left Namibian waters would return, hired a large helicopter from the private sector in South Africa and flew a contingent of volunteers from the newly formed Namibian Defence Force out to sea and arrested five Spanish Freezer trawlers at gunpoint. The vessels were ordered into port, the crews repatriated, the vessels impounded and the skippers and officers put on trial for illegal fishing. That action, and the subsequent arrest of three more vessels fishing illegally in Namibian waters in March 1991, and the heavy penalties imposed by the courts for their misdemeanours, put an end to the extensive fishing by unlicensed vessels in Namibian waters.

A significant lesson regarding enforcement is to be noted from the events leading up to these enforcement actions. Namibian companies that had been licensed to fish and were being charged fees for the quotas that they had been granted, resented the presence of unlicensed vessels that were fishing for free. They were also conscious that the dramatic reduction in the TAC was necessary in order to enhance the chance of recovery of the hake stock and that the illegal fishing was undermining this effort to the ultimate detriment of the Namibian industry. They thus provided detailed intelligence from the fishing grounds to the Namibian Government, without which the Namibian Government would have been unable to take enforcement action. Furthermore, they provided initial finance to the Ministry to hire the helicopter in order to better ensure secrecy, speed and the element of surprise that contributed to the successful implementation of the enforcement action.

---

6 A threat was made to bureaucratically block aid if "progress" was not made with a fisheries access agreement for EC vessels.

7 Initially under the Ministry of Agriculture, Fisheries, Water and Rural Development; subsequently a Ministry of Fisheries and Marine Resources was established.

Growth of the domestic hake fishery
Until independence the hake fishery was dominated by foreign fleets which landed the fish abroad for processing. The main market for Namibian hake was, and remains, Spain. There was a strong desire to build the domestic fishing industry both in the harvesting and the processing sub-sectors. An incentive structure was created to achieve this objective.

Incentives in the form of lower quota levies (see table 1) were created to encourage Namibian ownership and registration of vessels, the landing of fish for onshore processing, and the employment of Namibians at sea and on land.

It was also widely suggested by the Ministry of Fisheries and Marine Resources in the early 1990s that those who invested in onshore processing facilities could expect to be favoured in the allocation of quotas. Individual companies, when deciding on the future major capital expenditure, naturally assessed their expectations for the long term. The 1991 White Paper on fisheries policy, foresaw a total allowable catch of about 300 000 tonnes of hake once the stock was restored to a biomass that would maintain its maximum sustainable yield (Namibia 1990). In addition, individual fishing companies took an optimistic view as to what quota allocation they would receive. They relied on an expectation that, if they built a plant capable of handling a large quota, the MFMR would be morally obliged to allocate to them a large quota capable of keeping the plant fully utilised. What might have been individually rational for fishing companies proved not to be rational for the industry as a whole. By the mid-1990s, the processing capacity for white fish was about double what was available for onshore processing (Blatt, 1998). Potential resource rent was being dissipated on excess processing capacity. No estimates are available for the capital cost of this excess capacity.

Namibia's fisheries management system

In the immediate post-independence period, the Namibian Government worked systematically to establish an appropriate system of governance for the fisheries sector.

The new Government accepted assistance of the Fridtjof Nansen Programme with stock assessments which began in January 1990, just before independence (Sætersdal et al, 1999). These found that the hake stocks were depleted.

The process of delimiting and declaring an EEZ for Namibia[8] was undertaken as a matter of urgency, because the basis did not yet exist for legally ending open access to Namibia rich fishing grounds. The EEZ was formally declared less than three months after independence on 11th June 1990 with the adoption the Territorial Sea and Exclusive Economic Zone of Namibia Act (Namibia 1990).

---

8   Provided for by Article 75 of the UN Law of the Sea Convention, 1982.

Development of a fisheries policy

The formulation of a coherent fisheries policy was another urgent priority for the new Government. Namibia's fisheries policy can be traced from its constitutional roots, to the statement of those constitutional principles in the 1991 White Paper, "Towards the Responsible Development of the Fisheries Sector" (Namibia 1991), and then to its expression in the legislation which seeks to implement that policy.

The fisheries policy articulates two significant objectives. The first is to address effectively the depletion of several species which took place before independence and to rebuild the stocks "to their level of full potential" (ibid.). The second objective is to maximise benefits for Namibians from this sector, both in the harvesting of fish and in the processing industry. *Inter alia*, the policy also aims to encourage more employment of Namibians in both the fishing and processing industries and through the development of support and service industries (ibid). Notably, the White Paper also identifies a responsibility of the Government as to be "constantly assessing the social impact of resource exploitation, such as equity" (Ibid.).

Although often put less explicitly, a principal objective of Namibian fisheries policy is to use the opportunities presented by gaining control of the fishing industry, to address the extensive social and economic damage left by the pre-independence system of apartheid.

It is thus clear in the case of Namibia that fisheries resources are publicly owned and the resource rent associated with them, like that of mineral resources, belong to the people as a whole. There is also a clear commitment to achieve greater equity in the use of these resources.

Legislation

The articulation of a coherent fisheries policy laid the foundation for reviewing the existing South African legislation, the Sea Fisheries Act (1973), which remained the applicable legislation in Namibia. The South African legislation was repealed and replaced by the Sea Fisheries Act (Namibia, 1992). The Namibian fisheries management system is now based on the Marine Resources Act (Namibia 2000), which entered into force in August 2001 replacing the earlier post-independence legislation. The new Act maintains the system basically as it was under the Sea Fisheries Act (1992) but enhances it in certain respects.

There are two major areas of innovation in the new Act. Firstly, the new Act extends the powers to regulate fishing to Namibian registered vessels fishing on the high seas in order to enable the Namibian Government to exercise its responsibilities under the UN Fish Stocks Agreement (Namibia 2000, s37). The second major innovation is the creation of the Fisheries Observer Agency (ibid., Part IV). In the early 1990s the MFMR introduced an observer programme for fishing vessels. However, the observers were not formally employed by the Ministry and their employment and payment was too closely identified with the vessel whose fishing activities they were observing. The vessels owner paid a daily rate to auditors appointed by the Ministry who made payments to the observers for each trip undertaken. This tended to undermine the independence and authority of the observers.

The Marine Resources Act puts the position of observers onto an altogether more professional basis by creating the Observer Agency which is funded by Government through the Fisheries Observer Fund. Observers oversee and record the harvesting, handling and processing of fish at sea, collect and record biological data, and collect samples (Namibia 2000, s7(1)). Training programmes have been established, grades of fisheries observers defined and clear career paths developed.

Elements of the fisheries management system

A 'right of exploitation' is required to harvest each commercial species of fish or other living marine resource (Namibia, 2000, s32(1)). Previously rights were granted for periods of ten, seven and four years but, in June 2001when the new Marine Resources Act became applicable, the periods were changed to fifteen, ten and seven years and a new 20-year fishing right was added.

Total allowable catches (TACs), divisible into individual quotas, are set annually for eight species: hake, horse mackerel, orange roughy, alfonsino, pilchard, red crab, rock lobster and monk (the latter since 2001). Quotas may only be allocated to the holder of a right of exploitation.

Licenses are required for all vessels fishing in Namibian waters. Licenses are used to limit fishing effort in fisheries not subject of a TAC and quota allocation (eg the tuna fishery).

The basis for the length of time a right is granted are as follows:
- A 20 year right may be granted to a company that employs at least 5000 Namibians on land on a permanent basis.
- A 15 year right is granted to a rights-holder that is an enterprise at least 90% Namibian owned, with a "significant" investment in vessels or onshore processing facilities. 50% ownership of these inputs is regarded as significant.
- Ten year rights are granted to all other majority Namibian owned enterprises with at least a 50% interest in a vessel or onshore processing facility in the relevant fishery.
- Seven year rights are granted to enterprises that are majority Namibian owned but which do not have a 50% or greater ownership of a vessel or onshore processing plant in the fishery concerned.
- Note: Variations on these conditions exist relating to the size of the enterprise, the number of Namibians employed and on innovation. If a venture granted a seven or ten year right later fulfils the conditions for a longer term right, then that right may be extended on review by the Ministry of Fisheries and Marine Resources (MFMR). Similarly, if an enterprise no longer fulfils the criteria for which the right is granted, the right may be withdrawn or shortened.
- The structure of quota fees was established to encourage Namibian registration and ownership of fishing vessels. Categories are defined (Namibia 2001) as follows:
  - A Namibian vessel is one registered in Namibia, permanently based in Namibian waters, flies the Namibian flag and in which at Namibians

enjoy at least 51% beneficial ownership and whose crew is at least 85% (80% before 2001[9]) Namibian.
- A Namibian-based vessel is one registered in Namibia, permanently based in Namibian waters, flies the Namibian flag, has at least 51% beneficial Namibian ownership and a crew which includes Namibian citizens but of whom less than 85% are Namibian. It also includes foreign-flagged vessels with at least 85% Namibian crew[10].
- Foreign vessels are those that do not qualify as Namibian or Namibian based vessels.

The quota fees, charged per tonne, are based on these definitions of vessels (table 1). As a result of sharp rises in the Namibian dollar prices of hake products, quota fees were increased by 10% in May 1999 and by varying but substantial percentages in 2001:

Table 1: Hake quota fees applicable in the Namibian hake fisheries

	Hake quota fees per tonne in N$			
Period when fees where applicable	1994-1999	1999[1]-2001	2001[2]-	% increase 2001 ove 1999[3]
Freezer trawlers – foreign vessels	800	880	1450	64.8
Namibian based vessels	600	660	850	28.8
Namibian vessels	400	440	550	25.0
Wet fish trawlers – foreign	600	660	1200	81.8
Namibian based vessels	400	440	600	36.4
Namibian vessels	200	220	300	36.4

Notes: 1 - GRN 1999, 2 - GRN 2001, 3 - % increase on previous levy.

In addition, the following Marine Resources Fund levies, used for funding research and training, are charged per tonne of hake landed:

Table 2: Marine Resources Fund levies for hake

Whole fish (18.00)	N$ 22.50
Headed and gutted (25.00)	N$ 31.25
Fillets (45.00)	N$ 56.25
Broken/sour (25.00)	N$ 31.25

Notes: (GRN 2001a). Rates prior to the 2001 increase in brackets.

---

9 These definitions of vessels were different in some respects prior to 2001 (Namibia 1993).
10 Before 2001, a "Namibian-based vessel" could have less that 51% beneficial Namibian ownership and a crew of whom less than 80% were Namibian.

Fishing vessel licence fees are nominal at N$200pa for a vessels 200mt or greater. In addition, for each fishing vessel, used as a factory a fee is payable as follows:
- with a gross tonnage of or less than 4499 tonnes   N$ 20
- between 4500 and 8999 tonnes   N$ 500
- 9000 tonnes and more   N$ 1000

Marine Resources Advisory Council
Marine Resources Advisory Council provides a forum for broader consultation with stakeholders. It is made up on twelve people. The MFMR Permanent Secretary, who chairs the Council, is a member by statutory requirement. The remaining 11 members are appointed by the Minister. One other Council member is appointed from the staff of the Ministry. Five people who, in the opinion of the Minister, have "knowledge in matters relating to marine resources or any other expertise of relevance to the issues on which the Minister is required to consult the advisory council" are appointed. A further five "shall be persons who, in the opinion of the Minister, fairly represent the fishing industry or employees in the fishing industry" (section 25, GRN 2000[11]); the Minister must consult with any relevant trade association or trade union before appointing these latter five members of the Council.

The Minister exercises considerable powers under the Marine Resources Act, 2000 (Act No. 27 of 2000). However, the Advisory Council "shall advise the Minister in relation to any matter on which the Minister is required to consult the advisory council under this Act" Namibia 2000, section 24). The Minister must consider the best scientific evidence available when exercising his/her powers to set total allowable catches (TACs) but must also consult the Council before setting a TAC (sec 38). The Minister must consult the Council regarding the determination and imposition of fees payable in respect of harvesting marine resources. This includes the quota fees (sec. 44(1)), the Marine Resource Fund levies, and in the setting of the levies required to be paid into the Fisheries Observer Fund. The Advisory Council must also advise the Minister over "any matter which the Minister refers to the Advisory Council for investigation and advice" (sec.24).

The Minister of Fisheries and Marine Resources is required gain the approval of the Minister of Finance for the quota and effort related fees and for the Marine Resources Fund and Fisheries Observer Fund levies. Fees imposed on quota or effort are paid into general government revenue, while the Marine Resources Fund and Fisheries Observer Fund levies are paid respectively directly into those funds. The Permanent Secretary administers these two Funds with appropriate checks and balances and the Funds are subject to an annual audit by the Auditor-General (sec. 45, 46).

---

11 Marine Resources Act, 2000 (Act No. 27 of 2000) Government Gazette of the Republic of Namibia No. 2591, 1 August 2001, pp. 1-35.

Involving the industry in the science

The MFMR took the initiative to engage with the fishing industry over the scientific assessment of the status of stocks. In particular, the annual scientific meeting and other periodic meetings on the status of stocks and the condition of the marine environment were opened up to industry participation. Commercial vessels became involved in survey work leading to more active engagement by the industry in research activities at sea and generating a generally improved exchange between the Ministry and the industry. This appears to have greatly enhanced acceptance by industry of science-based decisions on TACs, but also greater debate over the reliability of the science.

KEY FEATURES OF SUCCESSFUL MANAGEMENT

Generally we judge success in relation to other experience. In other words, we use comparators, drawing conclusions by comparing one experience with another. Judging management of a fishery as "successful" does not mean that there is nothing that can be improved or that it is in some way being managed perfectly. Mistakes are made, particularly where there is a high degree of uncertainty, as there tends to be in fisheries. An assessment of success needs to capture the consequences of management activity over time and include the social, economic and biological dimensions of fisheries.

An economic and social success

Management of the Namibian hake fishery should be considered a success economically as sufficient resource rent is extracted to fully cover management costs and, in addition, make a net contribution to the national treasury.

The Namibian Government has succeeded in recovering the costs of management for most of the period since independence. This is an exceptional achievement in one sense, because so few governments, if any, have achieved quite the same. That said, the Namibian fisheries resources are very large in relation to population size[12], and the fisheries are relatively easy to monitor and control because there are only about 300 licensed fishing vessels in the whole of the Namibian fisheries sector and only two ports at which fish can be landed.

Each fishing entity, whether a company or individual, must have a "right of exploitation" granted by the MFMR to harvest Namibia's marine fisheries resources. Quotas are issued to rights-holders in eight fisheries, including the hake fishery. Officially quotas are not transferable, other than with the permission of the Minister. A levy is charged on the quota allocated to, and accepted by, the rights-holder, and must be paid regardless of whether the quota is caught. The levy charged is structured in such a way as to encourage the use of vessels that are Namibian owned and crewed (see table 1).

---

12 With a population of about 1.67million (2000), about 370kg was harvested per person in 2000.

Cost of management

The cost of managing the fisheries is summarised in Table 3. The figures include capital costs, which are spread over a ten year period (Wiium & Uulenga, 2002). The expenditure is financed from the central Government budget, the Marine Resources Fund and the Fisheries Observer Fund. (The fall in management costs in 1999 was due to the sale of the Ministry's helicopter and a patrol boat.)

Table 3: Cost of fisheries management in N$ '000s and as a percentage of landed value

	1994	1995	1996	1997	1998	1999
Monitoring, control, surveillance	24571	31524	45213	43456	48754	34455
Research	23026	17107	17201	23075	23623	22244
Administrative and other costs	4481	5688	6877	7372	9992	9258
Total	52078	54319	69291	73903	82369	65957
landed value*	881300	945500	1250100	1254100	1686500	1760800
management cost as % of landed value	5.9	5.7	5.5	5.9	4.9	3.7

Notes:. Capital costs spread over 10 year period. (adapted from Wiium & Uulenga, 2003)
* MFMR 1998, MFMR 1997.

The costs of management can be presented as a percentage of landed value (table 3, last row). However, in Namibia the landed value for hake largely reflects the transfer prices within large, vertically integrated companies and must be treated with caution. Landed value is not determined by an open market mechanism. The principal reason for the fall recorded in management costs as a percentage of landed value, is the sharp rise in the Namibian dollar valuation of the catch due to rising international prices coupled with devaluation of the Namibian dollar.

Throughout the post-independence period fisheries management has benefited from substantial donor funding. The data on donor contributions to the work of the MFMR, last published in the Ministry's Annual Report for 1998[13], shows that between 1995/6 and 1998/9 it ranged between N$30 million and N$ 36 million (MFMR, 1998b) per year. Wiium & Uulenga estimate donor contributions in 1996 as being "just above N$39 million" (Wiium & Uulenga, 2003). The reliability of these figures is questionable as difficulty assembling them has been experienced by the Ministry (ibid.). It is also not clear what contribution donor funding has made to the normal, essential management expenditure of the Ministry and to what extent it could be considered expenditure addressing the pre-independence failures of the apartheid system to provide adequately for the education and training of a large part of the population.

---

13 The Ministry has not published this data in its annual reports since the 1998 Report.

Cost recovery

Apart from normal company taxation[14], revenue is collected from the industry in the form of quota fees, bycatch fees, the Marine Resources Fund levies, the Fisheries Observer Fund levies and licence fees. The fees applicable to the hake fishery are described further in table 1 and table 2. The revenue actually collected by the Government is summarised in table 4 below. The landed value of the hake catch rose sharply as a result of both a steady rise in prices on the European market coupled with a decline of the Namibian dollar against, at the time, the Spanish peseta. Note that a mechanism has not been established to track prices and make the level of quota fees more responsive to changes in the value of production. Net revenue as a percentage of landed value dropped during the period in question but account is not taken in this table of the substantial increase in levies in 2001.

Table 4: Receipts received by Government from the fishing industry in N$ '000 (adapted from Wiium & Uulenga 2003) and net revenue as a percentage of landed value

	1994	1995	1996	1997	1998	1999
Quota fees	108600	90600	45500	72200	75200	91100
Bycatch fees	9600	8000	14800	5000	6200	9000
licence fees	30	162	162	158	160	172
Marine Resources Fund levies*	8600	7200	6100	8300	9900	13300
Fisheries Observer Fund levies*	5000	5131	5438	5371	5799	6026
Total	131830	111093	72000	91029	97259	119598
cost of management	52078	54319	69291	73903	82369	65957
net revenue	79752	56774	2709	17126	14890	53641
landed value	881300	945500	1250100	1254100	1686500	1760800
net revenue as % landed value	9.0	6.0	0.2	1.4	0.9	3.0

* These levies are paid directly into the Marine Resources Fund and the Fisheries Observer Fund respectively.

Figure 2 compares the landed value of the hake, the revenue collected from the sources listed in Table 4 above, and Government expenditure on fisheries management.

Revenue specifically from the hake fishery

The above discussion relates to costs of management and revenues for the fisheries sector as a whole. However, revenue collected from the hake fisheries constitutes a substantial proportion of the total revenue raised from fisheries.

---

14 Data on company taxation is regarded as confidential.

Figure 2: Expenditure on management and revenue raised in the Namibian fisheries sector compared to landed value for all species.

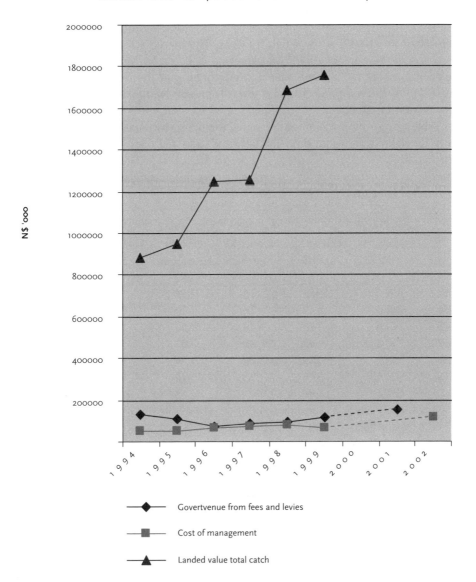

Notes: The sharp rise in landed value resulted from the dual effect of a rise in prices on the European market coupled with a decline in the exchange rate. Note that the cost of management rose to N$117m in 2002 (N$120m in 2003) (Namibian Economist 2002). An estimate of revenue for 2001 was N$190m, based on previous revenue levels and percentage increases in levies introduced in 2001.

The quota levy charged on Namibian hake remained static for most of the 1990s. There was a 10 percent increase in quota fees in 1999 and a further increase of varying percentages in 2001. By far the highest increases were for quota being utilised by foreign vessels, which rose by 64.8% for freezer trawlers in the hake sub-sector and 81.8% for wet fish trawlers. The increase for "Namibian based vessels" and "Namibian vessels" was less than half that for foreign vessels, as can be seen in Table 1. This indicates a strengthening of the incentive for fishing vessels to fly the Namibian flag and towards greater Namibian ownership[15] and crewing of the vessels.

Table 5 presents Government revenue, excluding corporate taxation, collected from the hake fisheries for the years between 1994 and 1999.

Table 5: Government revenue, excluding corporate taxation, collected from the hake fisheries 1994 –1999

	Revenue from all fisheries N$m	quota fees collected for hake N$m	fund levies collected for hake N$m	Total revenue from hake N$ millions	Revenue from hake as a % of all revenue	average hake quota levy paid per tonne N$
1994	126.8	68.482	3.1	71.58	56.5	457
1995	106.1	63.5	2.6	66.10	62.3	423
1996*	67.6	34.2	2.4	36.60	54.1	2011
1997	85.6	55	2.0	57.00	66.6	458
1998	88.9	52.3	2.7	55.00	61.9	317
1999	113.5	67.9	2.8	70.70	62.3	323
2000	96.8					

* Fishing was very poor in the hake fishery in 1996 due to the 'Bengeula Niño', which resulted in the catch being well below the TAC. For that year, as a concession to quota holders, the Government collected the fees on actual catch rather than quota allocated.

The data is not yet available that takes account of the 2001 increases in quota levies. However, as almost all vessels in the hake fishery now qualify as either Namibian or Namibian based, the increase in the revenue from quota levies can be expected to be about 32.5%. This assumes[16] that about half the vessels are classified as Namibian based vessels and half as Namibian vessels in the case of both freezer trawlers and wet fish trawlers (see Table 1). Note that the annual revenue from the hake fishery was between 56% to 66% of the total revenue from the fisheries sector between 1994 and 1999.

---

15 ...or apparent ownership! The real beneficial ownership of vessels is often not clear. Like fishing companies, ownership is camouflaged through layers of companies, and the use of nominee shareholders (Manning 2001).
16 Assumed, as this information is not available.

The Namibian Government has clearly done better than most other fishing nations in raising revenue from the fishing industry to cover the cost of management. The fees paid to government constitute recovery of part of the resource rent available in the fisheries.

POTENTIAL FOR COLLECTING ADDITIONAL RENT FROM THE HAKE FISHERY

If the resource belongs to the country as a whole, the resource rent, representing as it does the unexploited value of the resource, should accrue for the general good. However, it is notoriously difficult in practice to estimate rent with any precision, because of the difficulties of collecting the detailed data inputs needed to do so, because of the dynamic and variable nature of ecosystems and because of the variability of markets. In order to accommodate normal profits over time, it is either necessary to be able to keep abreast of market conditions in real time or, to allow for at least some rent to remain in the fishery, knowing that in less favourable times, lower profits will be earned resulting in something approximating normal profits *on average* over time.

There are indications, however, that considerable resource rent accrues to rights-holders within the Namibian fisheries sector. Data offering direct evidence of rent actually accruing in the Namibian hake fishery, (i.e. in addition to that collected by Government) is difficult to come by. However, one indication that rent is being generated is the evidence provided by the informal market for hake quota.

In perfect market conditions, the prices paid to lease quota for the year would reflect a discounted estimate of the resource rent expected from the harvesting of the quota during that year. Purchasers and sellers, with their intimate knowledge of the costs and revenues involved in the harvesting, processing and marketing of the fish, would reach a consensus on a price that still allows the buyer to make at least normal profits.

Some sketchy data on the prices paid for quota is presented in Table 6 below. It was collected by the author at different times from within the industry. The leasing out of quota is tolerated by the MFMR as a means of accumulating capital for investment in order to assist the new small quota holders, the "newcomers"[17], to become actively involved in the fishing industry. The MFMR does not approve of the practice if its purpose is simply to collect rent for the personal enrichment of the shareholders of the company owning the quota (Manning 1998). This results in a degree of secrecy regarding the prices paid and the conditions attached to the deals struck between parties. The data needs to be treated with some caution as the market for the hake quota is not a free and open market. In addition, the probable existence of excess capacity in vessels or processing capacity for some

---

17 This term is commonly used in Namibia to refer to the companies which had not been in the industry before independence but which have since been granted rights and quotas in the fisheries sector.

companies may mean that there is a tendency to consider only short-term operating costs (Manning 2001).

On the other hand, another important factor influencing prices paid for quota is the vulnerability to exploitation by larger companies of the relatively small quota holders, who do not own their own vessels but lease out the quota. Once a quota holder accepts a quota allocation, the quota fees must be paid regardless of whether the quota is caught or not.

Larger operating companies use the power they possess in fleet, processing and marketing capacity to offer prices well below what the quota might be worth in terms of the rent associated with it. An example relating to monk quota in 2001 serves as an example. Monk quota was leased out at the beginning of the period for N$ 4.50 per kg but, towards the end the year, quota left uncaught was given away by the small quota-holders to the large operators. They did this to avoid being liable for the quota fees but without receiving any income because the quota had not been caught. The general practice is for the quota fee to be paid by the operating company so that the price paid to lease the quota is net of all fees payable to Government (Manning 2003).

The differences in costs and revenues between frozen-at-sea and land-frozen product means that frozen-at-sea quota is considerably more valuable than land-frozen quota, despite higher quota fees being paid to the Government for frozen-at-sea quota.

Table 6: Informal market prices paid for quota allocated to be landed wet and for quota allocated for freezing at sea.

	prices paid for quota to be landed wet – N$/kg	prices for frozen-at-sea quota – N$/kg	average prices for landfrozen h&g product adjusted for nominal mass – N$/kg	average prices for FAS H&G product adjusted for nominal mass – N$/kg
1996	0.35	0.85		
1997			5.01	5.31
1998			6.44	6.89
1999			7.15	8.81
2000		2.20	7.23	10.36
2001	0.80	2.50	10.79	15.35
2002	1.20	3.00		

Notes: Average prices for land and sea frozen headed and gutted product, adjusted to reflect price for nominal catch (from Manning 2003).

The small quota holders, ie the newer companies which entered the industry after independence, and who are the sellers of quota, are in a weak bargaining position in this informal market. In contrast, the major operating companies, who exercise considerable power in fleet, processing and financial capacity, in market access and technical know-how, are in a position to extract the major part of the available resource rent (Manning 1998).[18]

The rent calculated using the price data from the informal market in Namibia for leasing quota can be expected to understate the rent actually accruing in the hake fishery, taking into consideration the above conditions. It is nevertheless instructive to examine what level of rent appears to be actually accruing in the industry using this data.

Table 7: Estimate of rent accruing to the Namibian hake fishery based on prices paid for the annual leasing of quota

	Wet-fish TAC mt	Frozen at sea TAC mt	Wet fish quota price N$/kg	FAS quota price N$/kg	Uncollected rent associated with wet fish quota N$m	Uncollected rent associated with FAS quota N$m	Total uncollected rent in hake fishery N$m	Total uncollected rent in hake fishery US$m[**]
2000	116000	78000	0.70*	2.20	81.344	171.147	252.491	32.37
2001	119000	81000	0.80	2.50	95.840	200.500	296.340	37.99
2002	117000	78000	1.20	3.00	140.400	234.000	374.400	48.00

* Estimate based on the wet fish quota price being 32% of FAS quota price as in 2001.
** June 2003 exchange rates.

On the basis of the prices paid for quota in Namibia, it is possible to argue that substantial rent accrues to companies in the hake fishery in Namibia, as indicated in the final two columns of Table 7. These sums may be regarded as conservative estimates of rent not collected by the Government and not dissipated in some other way. Government probably then collected about one quarter to one third of the rent actually available in the hake fishery in 2001.

BENEFICIARIES OF THE RENT ACCRUING TO THE INDUSTRY

Earlier extensive work analysed in some detail the degree of industrial concentration that has taken place in the Namibian fisheries sector since the implementation of the post independence fisheries policy in the early 1990s (Manning 2001). That analysis essentially provides a snapshot of the ownership and control of the hake industry in 1999/2000. The industry had consolidated around a handful of major companies. The 38 rights-holders in the hake fishery have increasingly been consolidated into groups of companies dominated by one or other of the big operating companies in the industry.

The major companies in 1999 received the benefit of about 80.3% of the total

---

18 The distribution of rent within the industry is discussed more fully in Manning 2001.

hake quota either directly or through though their share of joint ventures with smaller quota holders.

Those granted rights in the hake fishery fell basically into two groups. Those that were already operational in the fishing industry and those that were not operational but professed aspirations to become operational. In general, the first group were granted the largest quotas on the basis that they were already in possession of fishing and processing capacity, (in reality the existing industry) for which they needed quota. The second group, were granted significantly smaller quotas, and the rights granted were for shorter periods. The thinking of the MFMR was that these smaller companies should be given a chance to establish themselves in the industry for a period during which they could prove that they were serious about doing so.

It was clearly recognized that rent of considerable value remained available within the industry. This was not being taxed off in quota levies in order, it was argued, to give the chance to "newcomers" to invest in vessels and processing capacity, and become operational within the sector. In other words, the resource rent was to serve as a subsidy using the infant industry argument perhaps appropriate to part of the industry. However, by leaving the rent in the fishery (instead of taxing it off and providing explicit subsidies if it was felt that they were needed), the rent subsidy goes to all quota holders, theoretically in proportion to the share of the quota they received. This meant that the large, established and wealthy companies in the industry also received large resource rent subsidies (in proportion to size of the quota) in a country where there are far greater needs than to provide resource rent subsidies to the wealthy and often foreign interests.

The principal point to be drawn from this is that by far the greater proportion of the total quota allocated goes to the big players in the industry, and it is they to whom resource rent of considerable value mainly accrues, rather than the "newcomer" companies. Thus, from a policy perspective, it is clearly not the case that the "newcomers", the real targets of policy, are gaining appreciably from the quota allocations.

It would not make sense to undermine the industry by dismantling the present set of operational companies. However, in the context of development, it would make sense to tax off more of the available rent and, if it is considered desirable to grant subsidies to parts of the industry, to do so explicitly rather that give resource rent subsidies to the whole of the industry by leaving substantial rent available to all companies.

Rent, fishing effort and excess capacity

Reducing fishing effort is the key to generating rents in a fishery that has been overfished and employs excessive fishing capacity. To achieve this objective in the long term, the incentive structure needs to move away from the perverse incentive to harvest fish before others do so, and towards the use of just sufficient fishing effort to achieve optimal harvesting in the long term.

Effort was considerably reduced in the Namibian fisheries after independence, particularly in the hake fishery. In the preamble to a question put to the European Commission at the time of Namibia's independence, a Spanish MEP revealed that "more than 173 of the Community's freezer fishing fleet" were operating in Namibian water[19]. These vessels were targeting hake, Namibia's most valuable stock.

Following the declaration of an EEZ and the introduction of *rights*, there was a successful clampdown on fishing by unlicensed vessels and fishing effort was dramatically reduced. By 1999 there were 19 freezer trawlers, 15 longliners and 53 wetfish trawlers targeting hake, a total 87 vessels[20] mostly of smaller harvesting capacity than freezer trawlers.

Some evidence exists that there is still over-capacity in onshore processing and in fishing effort.

Using gross registered tonnes as a rough measure of effort, the catch per gross registered tonne (GRT) was 2.9 tonnes in 1994 and 3.4 tonnes in 1999. There might still be significant overcapacity in the hake fleet, however. We know, for example, from scrutiny of the Register of Fishing Vessels held at the MFMR, that NovaNam's two freezer trawlers, which together have a GRT of 3164 GRT in 1999, harvested the whole of NovaNam's freezer quota of 16364 tonnes, that is 5.2 tonnes per GRT. Based on this, the GRT of the fleet that would be needed to catch the whole freezer quota of 85 551 tonnes in 1999 would be 16 541 GRT. The GRT for the hake freezer fleet for 1999 was 24809 GRT. If all the vessels in the fleet operated as efficiently as NovaNam's, then 8269 GRT too much of fleet capacity was deployed, that is, about 50% more of what it would be needed. Gross registered tonnage, however, is only a rough measure of the fleet's harvesting capacity and many other factors contribute towards fishing effort, so this cannot be regarded as conclusive (Manning 2003). It does, however, raise a question that would require further investigation to answer.

While it might be the case that there still exists excess harvesting capacity in the fishery, the reduction of effort has been substantial and has resulted in an industry that is better able to cope with the ecosystem perturbations that result in variable catch limits being set from year to year.

During the 1990s, on-shore hake processing capacity developed way ahead of the hake catch being landed wet for on-shore processing. Individual companies made rational decisions to build plants capable of particular throughputs, and decided on throughputs higher than the quotas they were then receiving. They were encouraged to do so by expected increases in TACs and because the MFMR indicated that companies that did so could expect larger quota allocations (Manning 2003). While these decisions were rational for the individual company, they were not rational for the industry as a whole. By the late 1990s, onshore processing capacity for white fish was about 160 000 tonnes per annum (Blatt,

---

19 Question by Mr Areas Cantete on behalf of the EPP group, 6 April 1990. European Parliament Document B3-802/90.
20 Register of Fishing Vessels, Ministry of Fisheries and Marine Resources, Namibia.

1998). However, only 85000 tonnes of hake was allocated for onshore processing in 1998. These facilities are also utilised for the processing of other minor commercial species, but these come no where near absorbing the excess capacity.

Rent used as an employment subsidy

The principal reason that the Government insists on landing 60% of the hake catch wet for onshore processing is to maximise the number of jobs available in the fishing industry. The social desirability of this policy objective is obvious in a country where unemployment as a percentage of the economically active population runs at about 35% (Labour Force Survey 1997). It is understandable that the fishing industry, based as it is on Namibia's most valuable natural renewable resource, has been seen as providing a major opportunity for job creation. The importance of this to the Government is evident from the emphasis given to employment in the Annual Reports of the Ministry.

This policy, however, appears to involve the dissipation of potential resource rent of considerable value. Governments have every right to allocate resource rent for the purpose of achieving social objectives. It makes sense, nevertheless, for Governments to achieve such objectives in the most efficient way.

Fishing companies would prefer to receive the more lucrative freezer quota than the wet fish quota and would have been prepared to pay the additional N$220 per tonne, as it was in 1999. Cost in revenue not collected by the Government (by charging the wet fish quota levy and not that for freezer quota) in 1999 came to N$25.5million ($220^{21} 115920^{22}$) or N$10147 per extra job created.

Frozen at sea hake product attracts considerably higher prices than land frozen product (Manning 2003). The cost in lost revenue as a result of current level of land freezing should also be considered. The estimate of revenue lost in 1999 was about N$128.701million, or N$51214 per extra job created (Manning 2003). GDP per capita in Namibia is about N$12500 pa (National Accounts 1993-2000). Forfeiting these additional quota levies and loosing potential revenue on this scale is an expensive way of creating additional jobs. This policy needs careful re-evaluation. Instead of putting a resource rent subsidy into job creation in the fishing industry, consideration should be given to collecting up the additional rent through the quota levy and using it instead to subsidise jobs, for example, in the building of houses or schools. It should be possible in that way to create at least as many jobs as in the fishing industry but with the advantage of producing a better fish product that is more favoured by the market, and attracting better prices, and in addition have more houses or schools.

Although it might be argued that there is a positive socio-economic impact in creating additional jobs, this is being achieved at considerable cost in terms of fore-

---

20 See Table 1. Note that this figure rose to N$250 per tonne in 2001.
21 Wet fish portion of the TAC for 1999.

gone rents. This outcome is not optimal.

Despite these shortfall, the development of the fisheries sector in Namibia has had and very positive impact socially and economically in Namibia and the Namibian Government deserves credit for success in these dimensions of its management of the fisheries sector. This does not diminish the need to use even better what is widely recognised as Namibia's most valuable renewable natural resource.

Rights, and trade in rights

Once a right is created, it begins to gain a value in conditions where demand is greater than the capacity of the resource to supply. Where simply having a right has a value, rights-holders will find ways of trading those rights, even if the rights are officially not transferable. Thus, in the case of the Namibian hake fishery, rights-holders who had no capacity to harvest, process or market the hake, leased out the right or found ways of selling the right or a share of the right.

a. Rights are leased, most commonly under the guise of chartering vessels. A rights-holder, with little or no knowledge of the industry would "charter" a vessel to harvest the quota. Generally (in all cases I know of), this would be a vessel belonging to an operating company that has a quota in the fishery and is seeking additional quota to make the operation more efficient. The package deal would also involve the service of processing the catch and marketing it. The quota fee would be paid by the operating company. In fact, once the deal was done, the quota holder would have nothing more to do with the quota for the year, would bank the proceeds and enjoy the benefits. These small quota holding companies function simply as rent collectors. The "chartering" of a vessel in such cases is no different to leasing the quota (Manning 2001).

b. In many instances such arrangements as the above have been turned into permanent arrangements in the form of joint ventures. A web of joint ventures, an arrangement favoured by the MFMR, have sprung up around major operational companies. These tend to be arrangements whereby a "newcomer" quota holder gains by having an assured way of having the quota harvested and receive a portion of the resource rent accruing to the industry. The large operational company gains by acquiring more quota and, from evidence available, a substantial part of the resource rent (ibid.). Many of these "newcomers" are kept from becoming operationally involved and remain collectors of part of the available rent.

c. Selling quota also does not meet with the approval of Government, in as much as the right is not seen by Government as belonging to an individual to buy or sell. In most instances across all fisheries, with the exception of a handful of very small line and rock lobster rights-holders, the rights are held by companies. Many companies were established with the purpose of claiming a right and quota, and no doubt with the honourable intention of entering the fishing industry as productive, operational participants in the industry. Without any source of financing, most of these small "newcomer" companies became easy prey of the existing industry. The "newcomer" companies would tend to have no other assets other than a right and its associated quota. The financial institutions would not accept the quota as collateral because it was not transferable. The "newcomers", therefore, were forced to strike deals with the operating companies as the only way that they could have their quotas caught. If they did not do so they were still liable for payment of the quota fee and risked the loss of the quota. They had to bargain from a position of considerable weakness with the larger operational companies. Many of these companies sold shares to the larger operating companies. As the only asset the "newcomer" company had was a right and its associated quota, the bigger operational company in effect was buying part of that right and quota.

Rents have become, to some extent, capitalised into the value of the rights through being traded in these ways. The system, although described by the Government as a "individual (non-transferable) quota" system, has many of the characteristics of an ITQ system but without the formal acknowledgement of what it really is, and without a regulatory framework that deals with the realities of the system.

The established foreign interests and wealthy national companies found ways of protecting their interests in the Namibian fisheries sector utilising the skills of competent lawyers and accountants. The "Namibianisation" of the industry has not altered who the beneficiaries are nearly as much as might appear to be the case at first sight (Manning 2001).

Biological factors

Namibian fisheries management responsiveness to the state of the resource has been reasonably good over most the last decade, given the severe environmental perturbations and poor state of stocks at beginning of the period. It has demonstrated a capacity to cut TACs when the case for doing so is compelling.

However, in the mid-1990s, with the Benguela Niño experienced at the time, the biomass appeared to decline dramatically and the scientific advice from the Ministry's scientists was to cut the TAC. The MFMR, however, in consultation with the Advisory Council, continued to set catches considerably higher than this scientific recommendation.

Two factors played a part in this decision: firstly the considerable fleet and processing capacity that had developed and secondly, the doubts that were cast over the stock assessments as a result of the disagreements between the scientific consultants engaged by the industry and the Ministry's scientists regarding the status of the hake stocks.

As noted above, by the mid-1990s onshore hake processing facilities in Namibia had a capacity to process about 160,000 tonnes of hake per annum, about double that needed to process the hake available for onshore processing. The existence of this excess processing capacity, coupled with the fleet having grown in expectation of a linear rise in TAC, had generated great pressure on the MFMR not to reduce the TAC, particularly to the extent recommended by the Ministry's scientists.

The second factor relates to the genuine uncertainty that may surround the size of the biomass. This emerged through a debate between MFMR scientists, and scientific consultants employed by the industry regarding the size of the hake biomass. MFMR scientists argued that the fishable biomass in February 1996 was about 380 000 tonnes, a little improved on the previous year. This was based on sweep area surveys, acoustic surveys and analysis of catch and effort data from the commercial fleet. In contrast, consultants for the industry contended that the fishable biomass was in fact between 2 million and 3.5 million tonnes and that "TAC levels in excess of 200 000 tonnes are therefore advised" (OLRAC et al 1996, p5).

This enormous gap in the estimates of the biomass of the hake stocks constituted a serious disagreement between two groups of scientists using different stock assessment methodologies to draw their conclusions. (The industry scientists were using virtual population analysis (VPA) tools, relying on historical pre-independence catch data.)

The MFMR scientists responded to the industry by arguing that the surplus production model used by the industry could not produce acceptable results in this instance because it relied on dubious historical data produced by ICSEAF (Voges 1996) and contradicted evidence available from the sweep area surveys conducted by the MFMR and the Dr Fridtjof Nansen scientists.

These two widely differing views of the size of the fishable biomass were presented to the Fisheries Advisory Council which opted to recommend to the Minister that the hake TAC be set at 150 000 tonnes for 1996, twice the level of 75 000 tonnes recommended by the MFMR scientists for optimal stock recovery and 50 000 tonnes less than what the industry scientists argued would be a safe TAC. The challenge by the industry's scientists cast sufficient doubt on the recommendation of the Ministry's scientists (that the catch be drastically cut) for management to opt for the slight increase in the TAC instead of the decrease.

Management argued that they had to weigh up all the factors including the biological evidence, economic conditions, and social pressures over jobs and livelihoods. They also faced a yawning gap between the assessments of two reputable sets of scientific advisors and thus had a genuine question as to what the "best scientific evidence" actually was in this instance. Where there exists considerable doubt about the state of a resource, managers will tend to take an optimistic view of the state of the resource rather than risk what they see as almost certain serious economic damage being done to the industry.

(The doubts were exacerbated by a further puzzle that the scientists were pondering. The smaller stock of deep water hake, according to the sampling being undertaken of the catch, was being more heavily fished that was the Cape hake. Yet, in absolute terms, the biomass of the deep water hake was growing much more strongly than that of the Cape hake.)

It could be argued that the MFMR should have relied on the assessment of the Ministry's own scientific advisers in this instance, rather than those of the industry, which after all had a (short-term) incentive to keep their quotas relatively high. The decision arrived at underlines the fact that, in the face of high levels of overcapacity, fisheries managers tend more readily towards risk prone decisions.

On the whole, however, time and again the MFMR has demonstrated its readiness to make considerable cuts in catch in accordance with scientific recommendations and has thereby established a justified reputation for making responsible decisions on the basis of the best scientific evidence available.

Lessons drawn from the Namibian experience

Several lessons can be drawn from the experience of Namibia:
1. The Namibian rights-based fisheries management system offers a good example of fisheries where the incentive has shifted away from a race for fish to one where the fundamental incentive is to harvest as efficiently as possible. The result is that the fisheries generate resource rents of considerable value. Establishing clearly defined and enforceable rights created an incentive structure that contributed significantly to an alteration in behaviour of fishers which, in turn has lead to resource rent of significant value being realised. With everything to be gained by using vessels and gear as efficiently as possible, fishing companies have substantially reduced the size of the fleet.
2. Effective enforcement of rules is important in reassuring fishing companies that their rights are secure. Far higher levels of co-operation in the enforcement of rules is possible if the fishers appreciate the need for them and support their implementation than if there is an "us-and-them" atmosphere of conflict existing in the fishery.
3. In a rights-based management system, it is important to establish the principle that, because a right to benefit from the productivity of a publicly owned resource has real value, payment should be made for it. Payments should be introduced even when the resource is depleted and there is not much, if any, resource rent being generated. By establishing the practice early on, even if the payment of a fee or levy is token, it establishes the principle that the broader society has a right to benefit from the productivity of the natural capital of the country and that management of fishing activities is part of the cost of fishing. It also ensures that that portion of the rent does not become capitalised into the value of the right as a result of it being traded. The principle of payment for the use of the resource is not questioned in Namibia today, although there will always be efforts by companies to argue for lower levies to enable them to retain more of the profit for themselves.
4. Considering the influence that fleet and processing capacity had on management decision-making in the mid 1990s, the Namibian case underlines the fact that economic vulnerability arising from overcapacity gravely undermines the robustness of a management system. That can happen to the extent that even a fisheries authority that has shown a capacity for 'strong government' will not be able to act as robustly as might seem prudent, considering the state of the resource, for fear of seeing the economic collapse of the industry.

References

Anon. 1994: "Indian Ocean May Have El Niño of its Own", *EOS*, 75(50):585-586. American Geophysical Union, Washington.

Bianchi, G. et al. 1999: *Field Guide to the living marine resources of Namibia*.

Blatt, C. 1998: *Namibia Brief,* Hake Industry on Threshold of World Markets, No 20, January 1998.

Field, J. and Glazewski, J. 1992: "Marine Systems", in: *Environmental Management in South Africa*. R. Fuggle and M. Rabie. (eds), Juta & Co., Johannesburg.

ICJ. 1971: "Opinion on Namibia." *1971 ICJ Reports 27-58,* International Court of Justice.

Manning P.R. 1998: "Managing Namibia's Marine Fisheries: optimal resource use and national development objectives". PhD thesis, London School of Economic.

Manning, P., 2001, Review of the Distributive Aspects of Namibia's Fisheries Policy. Nepru Research Report No 21, Namibia Economic Policy Research Unit, 2001, Windhoek. ISSN1026-9231.

Manning, P. 2003: in: A. Arne, P. Manning and S.I. Steinshamn: *Assessment of the Economic Benefits African Countries Received From Their Marine Resources: Three Case Studies.* SNF-Report No. 05/03, Institute for Research in Economics and Business Administration, Bergen.

MFMR. 1997: Report of Activities and State of the Fisheries Sector 1997, Ministry of Fisheries and Marine Resources, Government of the Republic of Namibia.Windhoek.

MFMR. 1998: Report of Activities and State of the Fisheries Sector 1998, Ministry of Fisheries and Marine Resources, Government of the Republic of Namibia.Windhoek.

Namibia. 1990: Territorial Sea and Exclusive Economic Zone of Namibia Act 3, 1990.

Namibia. 1991: Territorial Sea and Exclusive Economic Zone of Namibia Amendment Act, No 30 *Gazette* No. 332.

Namibia. 1991: Towards Responsible Development of the Fisheries Sector, White Paper, Government of the Republic of Namibia.

Namibia. 1992: Sea Fisheries Act, No. 29 of 1992. *Gazette* No.493. 1 October 1992. Government of the Republic of Namibia.

Namibia. 1993: Sea Fisheries Regulations, *Government Gazette* No.566. 4 January 1993.

Namibia. 2000: The Marine Resources Act (Act No. 27 of 2000). *Government Gazette of the Republic of Namibia* No. 2591, 1 August 2001, pp. 1-35. Government of the Republic of Namibia. Windhoek.

Namibia. 2001: Government Notice No. 146. *Government Gazette of the Republic of Namibia* No.2579, Windhoek. 16 July 2001,.

Namibian Economist 2002: 18th April 2002, Fisheries ministry seeks close to N$120 million.

NEPRU 2002: Economic Review and Prospects 2000/2001, Namibian Economic Policy Research Unit, Namibia. At http://www.nepru.org.na/Annotations/NOT/NOT5.pdf.

O'Linn, B. and E. Twohig. 1992: The Commission of Enquiry into the procedures and practices applied in the allocation and utilisation of existing fishing rights. Report. Windhoek, Government of the Republic of Namibia. Windhoek. Government of the Republic of Namibia.

OLRAC cc and The Namibian Hake Association. 1996: Further scientific insight into the biomass and yield potential of the Namibian Hake Resource: implications for an in-season TAC revision in March 1996. Unpublished.

O'Toole, M. 1995: A preliminary report on the marine environmental conditions off Namibia from January to July 1994. Annual Research Meeting 21-22 February 1995, Ministry of Fisheries and Marine Resources., Swakopmund, MFMR.

Punt, A.E., D.S. Butterworth, A.J. Penney (1994): Stock Assessment and Risk Analysis for the South Atlantic Population of Albacore (Thunnus alalunga) for 1994. Unpublished.

Sætersdal, G., Bianchi, G., Strømme, T., Venema, S.C. 1999: The Dr. Fridtjof Nansen Programme 1975–1993. *FAO Fisheries Technical Paper* 391.

Shannon, L., A. Boyd, G.B. Brundrit, J. Taunton-Clark. 1986: "On the existence of an El Niño-type phenomenon in the Benguela System." *Journal of Marine Research.*

UNGA 1966. United Nations General Assembly Resolution 2145(XXI), 27 October 1966.

UNSC 1969. United Nations Security Council Resolutions 264 and 269, 1969.

Wiium, V.H. and A.S. Uulenga, 2003: Fishery Management Costs and Rent Extraction: The Case of Namibia, for inclusion in: *Costs of Marine Fisheries Management.* Ed. by W.E. Schrank, R. Hannesson and R. Arnason. (Manuscript supplied by author.)

The Australian Northern Prawn Fishery

Ian Cartwright[1]

Introduction

This study looks at the Australian Northern Prawn Fishery (NPF) and draws out the key features contributing to its relatively successful management over the past three decades.

The NPF is a limited entry, input controlled fishery targeting penaeid prawns. It covers an area of approximately 800,000 square kilometres in northern Australian waters and is focused on the Gulf of Carpentaria (Figure 1). The area fished is extremely remote from major cities and fishing vessels usually remain at sea for the whole season. Catches are transhipped to mother ships, which also provide supplies.

Figure 1: Area of the Northern Prawn Fishery

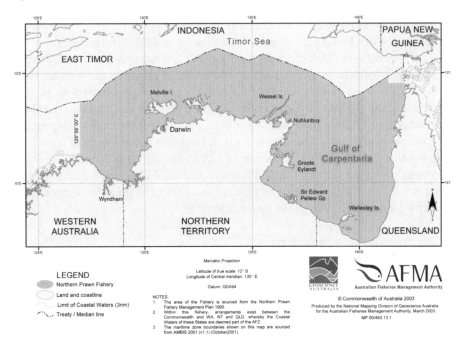

---

1 Thalassa Consulting Pty Ltd.

Using twin-rigged prawn trawlers, the Fishery lands an average annual catch of around 8,500 tonnes of prawns worth approximately $65[2] million, making the NPF Australia's most valuable Commonwealth[3]-managed fishery.

Background to the NPF

The Fishery
The NPF has two major components:
- A *banana prawn fishery*, which commences when the NPF season opens (1 April in 2003) and usually lasts for a few weeks. The fishery generally occurs during daylight hours and targets prawn aggregations, frequently using spotter aircraft. Fishing is very intense with vessels often working in close proximity and in strong competition. Very large catches can be taken in a short time.
- A *tiger prawn fishery*, which operates throughout the season, although most NPF vessels tend to switch to tiger prawns only when catch rates of banana prawns begin to decline. Fishing is generally undertaken at night and is more widespread across the NPF area than the banana prawn fishery.

Banana prawns were originally discovered in the Gulf of Carpentaria in the 1950s during exploratory fishing surveys undertaken to seek new grounds to relieve pressure on the Queensland East Coast and Northern New South Wales prawn fisheries. However, these surveys did not find commercial quantities of prawns. The discovery in 1965 of the first banana prawn school is attributed to CSIRO researchers, after two years of surveys in the Gulf of Carpentaria. CSIRO and industry then collaborated to learn more about the extent and nature of the prawn resource and a commercial prawn fishery was established in the Gulf in the late 1960s (Pownall 1994).

Initially, the NPF was a banana prawn fishery, with vessels targeting the abundant schools or 'boils' of prawns in the south-east corner of the Gulf of Carpentaria. The numbers of vessels rose dramatically in the early 1970s, partly as a result of huge catches in 1974 when in excess of 12,500 tonnes were landed. The open access nature of the Fishery, shipbuilding subsidies and Government development priorities for the Northern Territory resulted in a very rapid build-up of vessels and an expansion of effort across the area of the NPF.

---

2   All dollar values are in US$. Where conversions have been made, they have been based on 1A$ = 0.65US.
3   Fisheries in Australia are usually managed under either State or Commonwealth legislation, or, rarely, under joint jurisdiction. Generally, the Commonwealth has responsibility for fisheries beyond the 3 nautical mile limit of the Territorial Sea, except where agreement (under an Offshore Constitutional Settlement) has been reached with one or more States for a fishery to be managed under a single jurisdiction. This has been the case with the NPF since 1988, when agreement was reached with the relevant State and Territory governments to allow the Commonwealth to manage the NPF under a single jurisdiction. This agreement has allowed for more effective and flexible management of the NPF, particularly in sensitive nursery areas.

The banana prawn season shrank from a year-round fishery in the 1960s to a few months in the 1970s, to just a few weeks in the 1980s. In recent years, a poor banana prawn season, usually associated with reduced rainfall, may last little more than two weeks. This decline was exacerbated by a particularly dry decade in the 1980s that forced vessels to seek new fishing opportunities and led to an increase in effort on the tiger prawn fishery.

As the banana prawn fishery began to decline, attention turned increasingly towards tiger prawns. The tiger prawn fishery rapidly expanded until it too began to suffer from an excess of capacity and declining catches in the late 1970s.

The original vessels used in the NPF in the 1960s were of wooden construction and carried no refrigeration plant, relying on ice to refrigerate salt water (brine) tanks. Vessels used a single trawl net towed over the stern and spread by otter boards. The operation of these vessels was later assisted by the introduction of 'mother ships' in the 1960s, which allowed small vessels to spend more time on the fishing grounds and fish further from port. As occurred in other trawl fisheries, large, virtually self-contained steel freezer trawlers were introduced into the NPF in the 1970s replacing the brine and ice boats. These vessels, armed with new fishing technology, allowed the Fishery to expand rapidly. The Government permitted some foreign fishing and trawlers from Russia, Japan and Taiwan fished briefly in the NPF in the late 1960s and early 1970s.

Figure 2: NPF trawler and fishing gear[4]

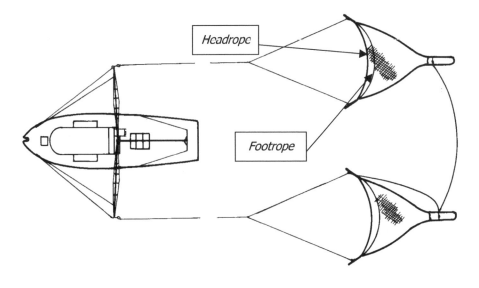

---

4 Diagram provided by the Fishing Technology Unit of the Australian Maritime College, Launceston, Tasmania, Australia.

There has been constant technological innovation among prawn trawler skippers and owners in the NPF as vessels have competed to maximise their efficiency and out catch their competitors. In addition to the switch to steel construction vessels and larger engines innovations included multiple gears (three or four nets per boat), improved propulsion systems (Kort nozzles and controlled-pitch propellers), hydraulic winches, electronic navigation and fish-finding aids and on-board processing equipment.

The current NPF fleet is made up of around 100 steel vessels, mostly between 18 and 24 metres in length. Nearly 70% of owners in the Fishery have entitlements to operate multiple trawlers. While the median age of the fleet is 17 years all NPF trawlers are maintained to a very high standard and equipped with modern electronics (echo sounders, sonars, global positioning systems and track plotters) and fishing gear. They use multi-rig prawn trawls and are restricted to two nets (twin-rig), with set headrope and footrope lengths (see Figure 2). All NPF trawlers are now required to carry an approved turtle excluder device (TED) and bycatch reduction device (BRD) (AFMA 2003).

A typical year in the NPF consists of two fishing seasons that have become shortened in recent years due to fishing pressure. The first ('Banana') season generally opens on April 1 when vessels target banana prawns – usually for three or four weeks or more in a good year. As banana prawn catch rates decline the NPF fleet progressively changes to tiger prawn fishing for the rest of the first season, which ends in May to early June. A mid year seasonal closure for the whole Fishery applies until August when the second ('Tiger') season commences. The second season usually closes in November.

There are two relatively distinct and somewhat polarised groups of operators in the NPF; the large, corporate fleet owners associated with Western Australia and those smaller Queensland east coast-based operators with single or two to three vessels. How these two groups view the Fishery and its management, and in particular the number of vessels that should operate, has significant ramifications for dynamics within industry consultative groups.

AFMA on behalf of the Australian Government has managed the Fishery since 1992. Prior to that AFMA's predecessor, the Australian Fisheries Service was responsible for management of the NPF.

BIOLOGY AND LIFE HISTORY OF TARGET SPECIES

The NPF Fishery is based on three major target species of prawns, the white banana prawn (*Fenneopenaeus merguiensis*) and the two tiger prawn species (grooved, *Penaeus semisulcatus* and brown, *Penaeus esculentis*). These species account for around 80% of the landed catch. The balance of the catch is made up of endeavour prawns (catches have recently increased to around 1000 tonnes per annum) and a number of minor prawn species[5]. Other marketable catch elements include small quantities of slipper lobsters or 'bugs' (Thenus spp.), scallops (Amusium spp.) and squid (Photololigo spp.). The biology of the three key prawn species strongly influences their management and fishing practices (BRS 2002).

The commercial prawn species in the NPF reach a commercial size after six months and can live up to two years. Growth rates vary considerably between species and sexes, with females generally growing faster and to a larger size than males. They are highly fecund, producing hundreds of thousands of eggs per female. These two factors are thought to have made them relatively resilient to fishing pressure, although this is moreso for banana than tiger prawns, as discussed elsewhere in this paper.

All species spawn offshore and after hatching larvae make their way inshore to the nursery grounds. Juvenile prawns live in the coastal and estuarine environment, favouring areas of seagrass beds or mangrove-lined creeks and rivers. After one to three months on the nursery beds they move offshore.

Banana prawns recruit to the fishing grounds in late spring and summer and form dense aggregations or 'boils' that are visible from the air[6] and on an echo sounder. The annual productivity and size of banana prawns has been linked to monsoonal rainfall levels that provide a salinity change to trigger migration offshore from estuaries as well as a physical flushing out of juveniles during heavy rain. The highly variable rainfall that is characteristic of the region is reflected in annual abundance, with annual catches of banana prawns varying between 2,200 and 6,600 tonnes over the last ten years.

Tiger prawns are active at night or during times of high turbidity. The tiger prawn fishing grounds are associated with areas offshore from extensive coastal seagrass beds, which are the preferred juvenile habitat. The two species (grooved and brown) of tiger prawns have differing life histories, which complicates stock assessment since they are not differentiated during commercial operations. The fishing grounds for tiger prawns are more extensive although in the last three to five the fishing area has contracted in response to the shortened seasons. Catches of both species of tiger prawns have varied between 2,100 and 4,100 tonnes over last ten years.

In recent years, depressed tiger prawn stocks have resulted in a growing emphasis on endeavour prawns, which have approximately half the economic value of tiger prawns. Whereas endeavour prawns have always been a byproduct of fishing for tiger prawns there are now reports of fishers increasingly targeting endeavour prawns when tiger prawns are difficult to find.

---

5   Other minor but commercially significant prawn species include: blue endeavour (Metapenaeus endeavouri), red endeavour (Metapenaeus ensis), red-legged banana prawn (Penaeus indicus), giant tiger (Penaeus monodon), western king (Penaeus latisulcatus) and red spot king prawns (Penaeus longistylus).
6   During the highly competitive banana prawn season vessel operators use light aircraft or 'spotter planes' to locate banana prawn boils and direct prawn trawlers to their location.

Stock assessment

The annual catch of the three key prawn species is generally made up of a single year class and considered to be heavily influenced by the numbers of prawns that successfully enter the fishery as a result of offshore post larval recruitment.

The NPF Fishery Assessment Group (NPFAG) has responsibility for assessing the dynamics and status of NPF species. Its membership and functions are discussed later. Stock assessments are based primarily on catch per unit effort (CPUE) data obtained from logbooks and limited survey information. NPFAG makes extensive use of models, particularly those relating to tiger prawns. An assessment strategy has been adopted of moving towards more realistic outcomes deriving from the use of stochastic forward projection models that allow consideration of uncertainty in future management decisions.

Banana prawns are considered to have been fully exploited since 1974, and annual catches seem to be a function of annual recruitment. As discussed, catch varies considerably, generally with rainfall although this relationship has been questioned in recent years when catches have been below those predicted by rainfall levels. While there does not seem to be recruitment overfishing of white banana prawns in the Gulf of Carpentaria this risk cannot be ruled out and a research project is underway to determine if this is the case. Similar work has been carried out on red-legged banana prawns and it does not appear that recruitment overfishing has been occurring. Recent and as yet unpublished research does shows a stock-recruitment relationship for banana prawns and this finding is likely to influence management decisions about this species in the future.

The AFMA Strategic Assessment document (AFMA, 2003)[7] notes that historical nominal (unstandardised) CPUE is constant but also notes that both species of tiger prawns are over-exploited, largely as a result of particularly high levels of effort in the early to mid 1980s. AFMA has concluded that the management responses, including the substantial effort reductions in late 1980s and 1990s, were inadequate to bring effort below the level of E_{MSY}.

Tiger prawns have been managed with the objective of achieving maximum sustainable yield (MSY). The stock assessment has approached this by attempting to determine the levels of spawning stock and effort that produces maximum yields – that is, S_{MSY} and E_{MSY} respectively and until 2001 these were the target reference points[8] for the Fishery.

Following a review of the NPF assessment (Deriso, 2001) S_{MSY} has now been confirmed by NORMAC as a limit reference point and not a target, and a more conservative target reference point has been established, such that there must be a >70% chance that the spawning stock biomass at the end of 2006 will be above or at spawning biomass targets.

The spawning biomass of brown tiger prawns is considered to be well below

---

7   See Section 6.8.
8   That is, the point at which remedial action was required.

S_{MSY} levels and it is likely that the other major tiger prawn species (grooved) is also below S_{MSY} but is not as depleted as the brown tiger stock.

The ecosystem

In common with other prawn trawl fisheries, the NPF interacts with the marine ecosystem primarily through the take of bycatch, the mortality of or interaction with threatened, endangered or protected species, and impacts on benthic communities.

Bycatch in the Fishery includes more than 450 vertebrate species (e.g. fish, turtles, sea snakes, sharks, rays and sawfish) and some 230 invertebrate species (e.g. crabs, squid and scallops). Prawn trawlers catch endangered turtles but TEDs have eliminated >95% of these interactions. Sharks and rays generally have a low ability to recover from fishing impacts (e.g. produce low numbers of young) and the effectiveness of TEDs and bycatch reduction devices (BRDs) varies for each species. In this respect sawfish (Pristidae) are a particular concern since they are vulnerable to capture, not easily excluded and prone to being heavily overfished[9]. Little is known about trawl impacts on sea snakes but indications are that some species may be at risk.

The disturbance and mortality of benthic communities as a result of interaction between the otterboards and the ground chain of trawl nets are also an issue in the Fishery. These impacts are mitigated by the facts that a relatively small proportion of the NPF area is trawled (around 14%) and the areas containing sensitive seagrass communities have been closed to trawling since 1983 (AFMA, 2003).

EVOLUTION OF MANAGEMENT ARRANGEMENTS IN THE NPF

The NPF has had a long history of management interventions to achieve set objectives. These interventions have been built around management plans, all of which has been conducted on a consultative basis with increasing industry involvement. Table 1 summarises this evolution.

Seasonal, area and time closures

Closures of one form or another have been a feature of the management arrangements for the NPF for most of its history. As the Fishery has evolved an increasing network of seasonal, time and area closures were established. These closures were implemented primarily to protect seagrass and other sensitive coastal habitats, which are significant nursery areas for juvenile prawns. Other closures have been developed to allow for the protection of small prawns to ensure they are at an acceptable size for harvesting, to permit pre-season gear trials and to maintain an adequate spawning biomass of prawns.

---

9 The 2002 IUCN Red List of Threatened Species lists five species of sawfish as being endangered and two others as critically endangered.

Table 1:   Evolution of management arrangements in the NPF

Year	Comments
Mid 1970s	Growing recognition that wide fluctuations in prices and resource availability caused economic instability in the NPF, which had in turn led to considerable overcapacity in the catching sector in years when prices or catches were low.
1977	Three-year interim Management Plan, introduced limited entry provisions, defined boundaries for the NPF[9], formalised seasonal closures, vessel replacement provisions and regulations on landings and processing.
1980	Management Plan containing most of the elements of the 1977 interim plan, but with a provision to continuously review closures and a liberal boat replacement policy that resulted in a continued increase in capacity entering the NPF.
1984	Revised management regime based on the 'unitisation' of the fleet into vessel-based units and licences to operate in the Fishery. Retention and expansion of system of closures and other input regulations.
1989	First Statutory Management Plan.
1995	Second Statutory Management Plan, reinforcing the property rights in the Fishery through the establishment of statutory fishing rights (SFRs).
2001	NPF Amendment Management Plan introduced, changing SFRs from vessel to gear-based units.

In 1971 and 1972 following concerns raised by industry, the Queensland and Commonwealth governments introduced seasonal closures in certain areas for banana prawns to protect pre-spawning adults. While these closures were initially made on biological grounds by 1973 it was recognised that the closures were more related to economic benefits due the high processing costs of small prawns (DPI 1982).

In 1987, in response to the effect of burgeoning effort on the tiger prawn stock and a CSIRO report that grooved tigers were overfished, a mid-year closure to protect spawning tiger prawns and a ban on daylight fishing were introduced alongside gear restrictions. The daylight ban was initiated by industry, which had reported large numbers of gravid females in the daylight catch (Pownall 1994).

More recently, fishing seasons have been reduced by lengthening the mid-year closure as one part of the strategy to reduce fishing effort on tiger prawns and rebuild their spawning biomass.

Fishing capacity, 'effort creep' and unitisation

The NPF fleet, like those in most other fisheries under input controls, has undergone a continuous process of adjustment. In 1977 when a moratorium on licenses (limited entry) was introduced around 150 boats regularly fished in the NPF[10]. The

---

9   Originally termed the 'Declared Management Zone' (DMZ).
10 Other vessels fished there but on an irregular basis and many Queensland east coast vessels would only go to the Gulf every 2-3 years.

criteria for a license was fairly liberal with even those who had been deckhands in the NPF and subsequently purchased boats on the Queensland east coast being eligible. In the end 292 vessels qualified, around twice the number than had previously fished (Jarrett 2001).

At the same time a ship-building subsidy[12] was in place that provided for a 'bounty' of 25% of the construction cost of a new vessel, including fishing gear. This subsidy, combined with generous taxation relief on the vessel cost, provided a major impetus to the construction of new vessels. Between 1977 and 1982 117 trawlers were constructed under the scheme. The generous boat replacement provisions of the interim management plan did nothing to stem the tide of investment or fishing effort in the NPF. In 1982, an NPF working group (DPI, 1982) had recognised that the most appropriate measure of fishing effort was swept area per unit time and that the effort control measures in place at the time were inadequate.

In 1985 following recognition that the boat replacement policy had been ineffective, a new system of license entitlements was developed. The two new entitlements were:
- Class-A units (vessel capacity) related to the sum of hull volume and engine power in kilowatts of an individual vessel (133,269 units); and
- Class-B units (a license to fish), which gave each holder the right to fish anywhere in the waters of the NPF[13] (292 units).

In recognition that excess capacity was a major problem in the NPF, the Commonwealth Government established an industry buy-back fund in 1985. Government provided funding of $1.95 million, and a further $3.25 million was provided to industry on a loan basis to be recovered from an industry levy. A voluntary buy-back scheme was started that proved to be of limited success, with less than 10% of the Class-A units in the fleet being removed by September 1986. In addition, 25 Class-B units, which were not purchased under the buy-back, were surrendered.

In 1986 CSIRO scientists provided convincing advice that the numbers of small tiger prawns entering the fishery had been drastically reduced and recruitment over-fishing was likely. A reduction in effort of 25% to protect sub-adult tiger prawns was suggested. In response, Government announced that a compulsory reduction of Class-A units to 70,000 (about one half of the original number) by December 1990 would occur. Industry appealed the decision and set about achieving the same target using a revamped buy-back system based on doubled levy payments and a range of other measures. These measures included:

---

12 This subsidy was devised so that Australian boat buyers could purchase locally at a price equivalent to that paid for vessels built overseas.
13 10 C-Class units were also issued for the Joseph Bonarparte Bay only (an area added to the original NPF area). The last of these units left the fishery in 1989.

- restriction to double gear (two nets per vessel rather than four) with maximum headrope lengths;
- additional closures;
- a ban on daylight trawling in the second part of the season; and
- a more restrictive boat replacement policy.

While some success was achieved (Class-A units were reduced by a further 17%, and Class-B units by 10%), unit prices rose steeply and the total cost of the reduction was more than $13 million (Meany 1993).

In 1990, and under a national policy imperative to address excess capacity across all Commonwealth fisheries, a further initiative to reduce capacity occurred and a new goal of 50,000 Class-A units by 1993 was agreed by NORMAC. The imperative to meet this goal was heightened by poor prawn prices and a deteriorating economic outlook for the Fishery. It was agreed that Government would put in an additional $3.25 million and guarantee a loan on commercial terms for up to another $26.6 million, provided the target of 50,000 Class-A units was reached, either under the buy-back, or if necessary by compulsory surrender. In the end, a total of about 27,800 Class-A units were purchased by April 1993 and there was a compulsory surrender of a further 18,374 units. This left 53,844 or 40% of the original 133,669 Class-A units and 137 or 47% of the original 292 Class-B units in the Fishery (Meany 1993)[14].

The capacity reduction process was far from smooth. There were some operators who saw the apparently draconian management measures as an investment that would result in longer term benefits for their operations and the Fishery. Other operators were reluctant to accept the reductions and fought the process all the way[15]. This reluctance led to major court battles, considerable angst within industry, and created a difficult climate for fisheries managers under which to operate. Some operators who found themselves with inadequate Class-A units could not fish until the requisite units had been purchased or modifications made to their vessels. This situation showed the inflexibility of vessel-based units and in the view of some laid the foundation for a search for a better management tool, resulting in the eventual adoption of gear SFRs.

Move from vessel to gear based units

In November 1999 there was a major shift in management arrangements from Class-A units based on vessel capacity to the use of gear entitlements. The latter provide for the establishment of an SFR in terms of the length of headrope (10cm)

---

14 The compulsory surrender did not achieve the target of 50,000 units due to agreed exemptions made to protect smaller boat owners and certain licence classes that were negotiated amongst Industry and government leading up to the final days of compulsory surrender.

15 One operator brought a case against the Government on the basis that the compulsory purchase was 'an acquisition of property on other than just terms'. The case was ultimately unsuccessful upon appeal.

and footrope (11cm) on the basis of greater flexibility in terms of matching fleet capacity to sustainable catches.

The transition to current arrangements was very contentious. In response to concerns by some operators from the NPF the Minister for Agriculture, Fisheries and Forestry requested that AFMA establish an independent allocation advisory panel (AAP) to examine issues related to the transition and in particular if there were grounds for 'transitional elements' that could allow for other than a one-for-one translation from Class-A vessel units to gear units (Commonwealth of Australia, 2000). The AAP endorsed the transitional arrangements from Class-A SFRs to gear SFRs on a one-for one basis, as judged against the criteria (i.e., AFMA's legislative objectives) of cost effectiveness, ecological sustainability, economic efficiency, accountability and cost recovery (SRRATLC, 2001).

The Queensland NPF Trawler Owners' Association (NPF (Qld) TA) and others lobbied against the move towards gear units and a Senate Inquiry was established[16]. The Inquiry heard a wide range of evidence contained in the 80 submissions, the most significant points of which can be summarised as follows:

- the CSIRO considered that the tiger prawn stock had declined significantly since the early 1980s, raising concern for the continued viability of the Fishery, and recommended that there be a 35% reduction in fishing effort for tiger prawns;
- the NPF (Qld) TA and others argued that the stock decline was partly a function of an over-estimate of 'effort creep' within the CSIRO assessment, and that declines partly reflected variations in patterns of catching effort resulting from changes in fleet size and fishing strategies;
- that the proposed use of headline length as the basis for gear units to control fishing performance would be considerably less efficient than alternatives[17], which would more effectively control swept (trawled) area performance; and
- the NPF (Qld) TA argued that the one-for-one translation between vessel-based units to gear units was 'grossly unfair and inequitable for small operators'.

The focus of much of the debate at the Inquiry was on two issues.

First, concern was raised over the method used to allow for effort creep[18] in the NPF and how it had been applied to the NPF stock assessment. In the past, CSIRO had applied an assumption that the fishing power of the NPF fleet had been increasing on average by 5% per annum. The counter argument was that this esti-

---

16 The Inquiry was conducted by Senate Rural and Regional Affairs and Transport Legislation Committee (SRRATLC) on the Northern Prawn Fishery Amendment Management Plan.
17 These alternatives included additional seasonal closures, another buyback programme, policing engine power, time units (e.g. nights fishing) and other net restrictions.
18 Changes in the catchability coefficient, in the catch/effort relationship that forms the basis of most stock assessment models (Y=q.f. B, where Y = yield (or catch), q = catchability coefficient, f = nominal or measured effort and B = stock biomass), due to the performance of fishing vessels, efficiency of trawl gear and the abilities of the skipper. Other changes in q may depend on *inter alia*, prawn behaviour, biomass levels and fleet size.

mate, while fitting the stock assessment model, did not reflect the realities within the Fishery and could make the stock assessment appear overly pessimistic. Further, it was suggested that it would be possible, using an engineering approach, to calculate the impact of key inputs over time and thereby more effectively standardise effort.

Secondly, the means of allocating gear units (headline length) in direct proportion to Class-A units as a means of allocating and controlling effort was questioned. Some operators, who were predominately from the Queensland east coast, argued that the directly proportional allocation of gear SFRs for boat unit SFRs would favour larger vessels, which would be allocated more headline length units than they were then currently using. With the removal of vessel constraints, this would enable these larger boats to increase horsepower (and swept area) over time. The smaller vessels would be left with inadequate headline length and would have to purchase headline units to remain viable.

A second group of operators, which were generally fleet owners using larger vessels from ports other than the Queensland coast, argued that the SFRs in the Fishery were the 'currency' of right in the Fishery since 1985. On this basis they considered that it was neither equitable nor acceptable to vary allocations other than directly in proportion to the rights already held. This group also considered that catching performance was more closely related to gear size than it was to hull size and engine horsepower.

A number of other issues were raised including the likely inefficiency of using headline length as the key input control over the use of engine power or thrust of the vessel (related to the second point above).

Notwithstanding the arguments raised, the report of the Inquiry supported the introduction of the new management plan, including the one-for-one transfer from Class-A to gear SFR arrangements, '... as the most equitable and legally defensible means of achieving a long-term sustainable northern prawn fishery' (SRRATLC, 2000).

In the opinion of some researchers, cuts in capacity in the NPF have been somewhat piecemeal and inadequate to deal with the ongoing creep that occurs between the call for capacity cuts and implementation of the necessary management changes. There is little doubt that reaction time between stock concerns and capacity reduction was slow in the intervening period between the compulsory reduction in capacity in 1993 and the introduction of gear units in 2000[19]. Reaching agreement on gear units was a difficult and protracted process, with strongly divided opinions on the need for, and preferred method of, capacity reduction. NORMAC was keen to avoid another disruptive compulsory surrender of units and in the spirit of co-management worked to reach an outcome that was acceptable to the majority of industry, which took time. Since the change to gear units, there is a strong record of NORMAC agreeing to action in response to stock assessment advice.

---

19 The move to gear units was first recommended in 1992 by the NORMAC Working Group on Future Arrangements. It took until 2000 before gear units were implemented in the NPF.

Property or harvesting rights

Property rights originated in the NPF in 1977 when a total of 292 endorsements[20] were issued as part of the interim management plan that introduced a limited entry policy and created a three-year moratorium on new licenses.

Following unitisation in 1984, harvest rights were allocated to the NPF fleet on the basis of Class-A (capacity) and Class-B units (license to fish). These rights were transferable and allowed fishers to arrange vessel-related inputs within a fixed cap of Class-A units for the Fishery as a whole. The units:
- rapidly acquired a value as they were traded between operators, eventually becoming the currency for the Fishery;
- were used as collateral on which to obtain finance; and
- were the mechanism by which adjustment of fishing capacity occurred.

The unit buyback schemes and compulsory surrenders during the 1980s and early 1990s outlined earlier reduced the numbers of Class-A and B units, increasing their value but not substantially changing the nature of the rights (Jarrett 2001).

In 1994 property rights in the Fishery were strengthened through the establishment of SFRs based on Class-A and B units, with the former being changed in 2000 from vessel to gear-based SFRs.

Industry consultation and participation

There has been a long history of consultation in the NPF, initially between governments with limited industry involvement. As the Fishery has developed industry has become more involved in the management process, initially through the Northern Fisheries Committee and since 1984 through NORMAC.

From 1983 to 1995 CSIRO, with the assistance of Commonwealth funding, provided a full-time liaison officer whose major task was to provide a face to face link between researchers and prawn fishers. This was achieved by a programme of port visits and the production of a wide range of 'plain English' pamphlets explaining the results of research and how industry could contribute to the research effort. AFMA and its forerunner, the Australian Fisheries Service, provided port-based logbook officers to the NPF, commencing in the mid 1980s. These officers saw each operator around five times a year, either on port visits or by going to sea with motherships. As logbook compliance improved the need for these officers to go into the field reduced and finally ceased in 2001.

Relationship-building between industry, managers and researchers was pivotal in gaining the cooperation and trust of industry in the completion of logbooks and obtaining voluntary assistance with sensitive information, including data on trawl tracks and changes to fishing gear and rigging.

---

20 Not including 10 Class-C units (see footnote 13).

Current management arrangements in the NPF

Policy and legislative frameworks

The NPF is managed by AFMA under a statutory authority framework model. This model features a high level of autonomy from the Minister for day-to-day decision-making, an expertise-based Board, rights-based management and a strong partnership approach with key stakeholders (see Figure 3).

Figure 3: AFMA Organisational Structure

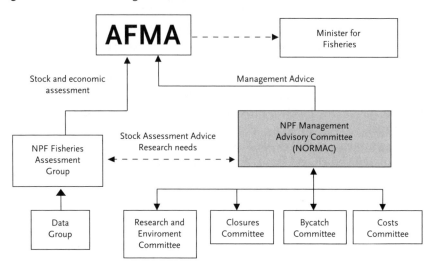

The eight member expertise-based Board of Directors is responsible for overseeing AFMA's operations and making high-level decisions on fisheries management matters. In the case of the NPF the AFMA Board receives advice from two main areas, the FAG and NORMAC.

The NPFAG has a membership drawn from AFMA, science (CSIRO and a second independent scientist[21]) and economic specialists, industry and conservation NGOs. The task of the FAG is to provide advice on biological and economic issues impacting the Fishery and to undertake an annual fishery assessment. The FAG reports both directly to the Board and through NORMAC.

In 2003, NORMAC had nine members, comprising an independent Chairperson, five industry members, an AFMA member, a research member, a conservation NGO member and a permanent observer representing the State and Territory governments. The five industry members are selected for their knowledge and expertise in the Fishery. NORMAC's major function is to act as AFMA's main point of contact for the NPF, providing the forum for discussion of key issues relating to

---

21 This scientist is independent of CSIRO and provides what is seen by Industry members as a 'check' against what has been viewed as a monopoly on science in the NPF by/for CSIRO.

the Fishery. NORMAC provides direct advice to the AFMA Board on issues such as fisheries management arrangements, scientific, environmental and economic research, compliance, monitoring and budgeting.

In managing Commonwealth fisheries resources for the benefit of the community as a whole AFMA is bound under its legislation to pursue the objectives of:
- efficient and cost effective management;
- ensuring fisheries are conducted in a manner consistent with the principles of ecologically sustainable development (ESD) and the precautionary approach;
- maximising economic efficiency;
- ensuring accountability to the Australian community; and
- achieving Government targets relating to cost recovery.

AFMA's management policies are based on the pursuit of efficient harvesting regimes and providing positive, market-based incentives for commercial fishers to enhance and conserve resources. A key mechanism to achieve these outcomes is the provision of secure and transferable access rights (that is, SFRs) within statutory management plans that provide a stable legislative environment. Fisheries policies can also be pursued through binding conditions on either SFRs or fishing permits, thereby permitting a level of necessary flexibility in a dynamic fisheries environment[22].

Management measures

Effort control in the NPF is primarily through limited entry and gear restrictions. Two classes of SFRs directly related to these measures are held in the NPF, and in 2003 these were:
- 102 Class-B SFRs each entitling the holder to operate in the NPF;
- 53,844 gear SFRs each representing 7.5cm of headrope and 8.625 of footrope length.[23]

Under the NPF management plan a trawler used by a concession holder must be nominated against one Class-B SFR and a threshold number of gear SFRs[24]. Other key management arrangements in place under Fishery Directions for the 2003 season are:
- no more than two prawn trawl nets with size restrictions relating to the number of gear SFRs held, and a requirement to have them measured by a nominated third party and permanently tagged;
- one 'try'[25] net of restricted dimensions (with a maximum headline length of 3.66m) per vessel;

---

22 Or by binding conditions on three monthly scientific permits.
23 Every net used in the NPF has to be measured by a designated officer and officially tagged – this tag must remain on the net at all times.
24 100 in 2003.
25 A small sampling net used to locate high densities of tiger prawns.

- various seasonal, area and time closures, with two major fishing seasons;
- a prohibition on daylight trawling during the tiger prawn season;
- pre-season vessel movement bans in areas where banana prawns may aggregate to ensure compliance with the start of the season;
- bycatch restrictions (mainly of fish, in terms of number and species, including prohibitions on landings); and
- compulsory use of turtle excluder and fish excluder devices (TEDs and BRDs) or a modified TED.

The last round of effort reduction through net and season reductions is now in place and the effects on the Fishery are being closely monitored.

It is worth noting that formerly it was necessary to go to the Australian Parliament and table amendments to the NPF management plan, including those relating to capacity reductions. This could take as long as six to 12 months and involve public consultation. Under a revision to the management plan[26], it is now possible to change the pool of gear on which gear SFRs are based through an administrative process that can be implemented within 24 hours. However in the case of changes to gear length a four-month minimum period is allowed to allow industry time to adjust. This is a significant step in allowing adaptive changes to management to occur.

Statutory fishing rights

As with other Commonwealth fisheries, the NPF has strong fishing rights. These are based on the Class-B and gear SFRs discussed earlier. There is a general requirement for Class-B SFR holders to have a minimum holding of 100 gear SFRs to operate in the Fishery and, once the holding drops below this, the Class-B SFR is automatically cancelled. At a value of over $7,000 per gear SFR, the holding of a Class-B SFR and not fishing becomes a very costly exercise in terms of asset return and the management levy payable on the 100 gear SFRs. The minimum holding requirement provides a mechanism to gradually reduce numbers of boats in the Fishery as a result of trade in gear SFRs.

Cost recovery

The current Government policy on cost recovery originated in the mid-1980s as part of the general philosophy that the beneficiaries of Australian Government services should meet the cost of those services in accordance with the concept of user pays. Along with the move to implement cost recovery was recognition that fishing operators were entitled to have a significant input to fisheries management decisions, including those which directly affect management costs.

---

26 By changing from each SFR being an absolute amount of headline length (e.g.10cm), to a share of the pool of total allowable headline length gear units.

In accordance with Government policy the NPF industry pays for costs directly attributed[27] to fishing activity on a full cost recovery basis, with Government paying for activities that may benefit the broader community as well as the industry. Recoverable management costs include the running costs of NORMAC, part of the costs of the NPFAG, licensing, AFMA's day-to-day fisheries management activities, the cost of developing and maintaining the management plan, and logbooks and surveillance - but not enforcement. In addition, there is a government-imposed compulsory minimum levy of 0.25% of the gross value of production (usually $160-200,000) of the Fishery payable to the Fisheries Research and Development Corporation (FRDC). These funds are matched by government and are available for FRDC-Board approved research projects. Industry also provides additional voluntary funding for management-related research priorities – this funding totalled close to $300,000 in 2002-2003.

In the 2002-2003 budget year around 40% of AFMA's total budget of $17.7 million was met by industry levies and license fees across all fisheries, with the balance coming from Government appropriations and other revenue. In the case of the NPF, in 2003 of a total budget of $1.25 million, industry met 74%, or $0.92 million.

THE NATURE OF MANAGEMENT SUCCESS IN THE NPF

Ability to address sustainability issues

On face value the current status of prawn stocks in the NPF alone may not be a good indicator of management success. Banana prawn stocks are considered to be fully fished and 'probably sustainable' at current levels and both main tiger prawn stocks (brown and grooved) are considered to have been overfished since 1997 (BRS 2002). The complex nature of the stock assessment, due particularly to the uncertainty of some aspects of logbook data and the effect of fishing power, has made interpretation of data difficult, leading to uncertainty about appropriate management measures.

The strength of fisheries management in the NPF has however been the ability of the Fishery to identify sustainability challenges and respond by putting management measures in place to address them. Probably the strongest evidence of this ability has been the actions taken over time to reign in effort towards the level necessary, as indicated by stock assessments.

The fleet reduction in the NPF fleet from 302[28] vessels to approximately 96 in

---

27 In the case of the development and maintenance of management plans, for instance, cost would be attributable to commercial fishers on the basis that, in the absence of commercial fishing, no management plans would be required. Costs are also fully recoverable from commercial fishers on the basis that the benefits of the introduction and maintenance of management plans flow predominantly to the fishing industry.
28 Includes 10 Class-C licences (see footnote 13).

2003 has been achieved since 1984, in addition to a complex range of gear restrictions and closures. Upon the issue of gear SFRs a 15% reduction was implemented as an initial response to rebuilding tiger prawn stocks[29]. Following an independent review of the stock assessment in July 2001, based on a NORMAC recommendation AFMA agreed to measures to reduce the effort on tiger prawns by 40%. This was achieved by a further 25% reduction in gear units and a 13% reduction in the length of the fishing seasons. The target under these new effort reductions is to rebuild the tiger prawn stock to S_{MSY}[30] within five years (by the end of 2006). Progress towards this goal will be reported to all stakeholders through the stock assessment and AFMA's accountability provisions.

AFMA has stated[31] that it '...has high confidence that .... the broader marine environment is being managed in a sustainable and precautionary manner'. In making this statement, AFMA notes that:
- an investment of over $2.3 million of research and industry funds have been spent on the development and implementation of bycatch reduction strategies;
- bycatch reduction strategies include the use of TEDs and BRDs, area closures and the ban on the take and retention of sharks, rays, sawfish or their products; and
- in 2003, an annual independent data collection program will commence.

The NPF has been sensitive to the need to demonstrate environmentally sustainable operations within the Fishery. TEDs and BRDs have been shown to be particularly successful in reducing the mortality of turtles and preliminary estimates are that turtle bycatches have been reduced by up to 99%, small fish by 8%, and sea snakes by 12%. The bycatch of stingrays and other fish has been reduced substantially. This has occurred at the cost of a loss of prawn catches of around 6% (CSIRO data[32]), and a substantial increase in the quality of prawns and, as a result, better prices. By fitting TEDs, market access to the US market was also achieved.

The current system of area closures provides protection from trawling to sensitive seagrass and other coastal habitat in the NPF. In 2000, the last year for which data is available, around 811 of 7281 grids (6 minutes x 6 minutes) were fished and around 6% of the total area is seasonally closed and 2.1% (AFMA 2003) permanently closed. While the number of closed areas in the NPF provides a degree of confidence that action is being taken to address biodiversity and conservation issues the results of current research will confirm what, if any, further action needs to be taken.

---

29 In addition 512 temporary 'top up' gear units were issued to vessels with less that 300 Class-A units to provide an opportunity for them to adjust to the new management arrangements. These units expired in August 2002.
30 SMSY being the level of spawning stock that, if present, would lead to the production of the Maximum Sustainable Yield.
31 The NPF Strategic Assessment Report (AFMA 2003). See also Section 6.8.
32 Personal communication, David Brewer, CSIRO, Cleveland.

A restructured fleet

The process of fleet restructuring and capacity reduction has been a continuous one in the NPF, as outlined earlier. The series of industry-funded (with limited government assistance) buybacks in the 1980s and early 1990s was summed up in 1994 by an ex-AFMA senior manager Frank Meany thus:

> *For all its imperfections the achievements of the NPF restructuring speak for themselves. In a period of seven and a half years the size of the NPF fleet was reduced from a maximum of 302 boats to a maximum of 137. This is believed to be the most significant restructuring of a viable fishery achieved anywhere in the world.*

A number of lessons were learned from the buy-back (Meany 1993):

- during restructuring, the price of fishing rights may not reflect a natural level (that is, one that reflects its true earning capacity) and may become artificial, eventually representing no more than the price that the buyer is prepared to pay;
- major restructuring should be done over a long enough period such that it is possible to buy out capacity during periods of poor returns and profitability, when there are more sellers than buyers;
- in the absence of a market for boats withdrawn from the fishery, it will be cheaper to focus on boats near the end of their operational life, unused capacity or obsolete boats;
- the buyback should operate at arms-length from the fishery; the NPF scheme was run by industry, and information on prices for units became well-known, resulting in a ratcheting up of prices[33]; and
- a phased compulsory surrender of units may have reduced the cost to industry (and government) of restructuring – however it may have meant that individual operators who could not purchase units would have had to cease fishing.

Generally the larger operators will be happy to see reductions in capacity (the gear pool) since they can remain in the Fishery with fewer operators overall. The small operators would generally prefer the longer closures as an alternative to gear reductions since they can remain in the Fishery and the number of operators will be kept higher. Larger operators are generally opposed to any lengthening of closures, and indeed some have argued for current closures to be reduced, on the basis that it is inefficient for vessels to remain idle for a substantial period of the year.

Fleet restructuring has occurred at a pace that has allowed the maintenance of a reasonable level of profitability in the Fishery over time, such that most opera-

---

33 Even working outside industry, a government-sponsored buyback will set the absolute base price for sale of any units regardless of how ludicrous it may be commercially (i.e. it could be twice as high as normal commercial rates would dictate, and no-one will accept anything less, hence it becomes a base price, and soon becomes over-run by the commercial trading levels).

tors have been financially able to adjust to capacity reductions. The value of the SFRs in the Fishery, reflecting a general level of confidence in the sustainability of catches and profits, has provided marginal operators with a financial incentive to sell up and leave the Fishery.

Generation of net returns and industry profitability

Since it is not considered possible to calculate resource rent[34] a proxy measure of net returns[35] to the Fishery is used in government assessment of the economic performance of the Fishery (ABARE 2002). Net economic returns from the Fishery have ranged from a high of $27 million in 1994-1995 to a low of $5.4 million in 1991-1992. The net economic returns for 1990-1991 to 1999-2000 are shown in Table 2 below. The Table is illustrative of two major sources of variation:
- export markets for prawns and currency fluctuations (around 90% of the NPF catch is exported); and
- environmental factors including intensity of the monsoon season, resulting in varying quantities of prawns available for harvest.

It has been suggested (Brown et al, 2002) that owing to the uncertainty of the stock assessment some of the net returns include return from the liquidation of part of the prawn stock, in which case these the level of these returns may not be sustainable[36].

The financial performance of vessels is also monitored by ABARE. The estimated return to boat capital decreased on average for the fleet from 20.7% in 1998-99 to 14.4% in 1999-2000. The estimated return to full equity (including the value of SFRs and licenses) was 4.3%, down from an estimated 8% in 1997-8. (Brown *et al*, 2002). There is little doubt among industry that the next few years will be tough as efforts are made to rebuild stocks of tiger prawns.

Much of the rent from the NPF has been capitalised into license values, and the total value of NPF gear SFRs have been estimated as approximately $350 million.

a Amount attributable to fishery. b Cash costs include imputed operator and family labor costs but exclude license and levy payments and management costs. c Replacement capital (depreciated capital for 1990-91 and 1991-92) is calculated by applying the replacement capital value for boats in the three size groups in 1992-93 to the population in 1990-91 and 1991-92 and then adding depreciation estimates. d Net return measure does not include any management costs. e Calculated

---

34 The long-run excess of income from a fishery over the fishing and management costs.
35 The measure of net returns to the fishery includes only those receipts and costs that are attributable to the NPF and do not include those while fishing in other related prawn fisheries.
36 It should be noted that when the stock is at a level that is below S_{MSY} and the effort is higher than F_{MSY}, the situation is sustainable and in equilibrium, but it is not the maximum that can be sustained

Table 2: Net returns to the Northern Prawn Fishery in 2000-2001 dollars

	Revenue (a)		Cash Costs (a,b)		Capital (a, c)		Net returns (d, e)		Mgt costs (f)		Vessel numbers
	$m		$m		$m		$m		$m		
1990-91	91.39	(2)	67.7	(2)	58.2	(1)	12.4	(13)	na		169
1991-92	71.04	(3)	57.9	(3)	47.2	(2)	5.4	(25)	na		160
1992-93	85.8	(6)	66.5	(5)	46.2	(5)	13.5	(15)	na		129
1993-94	94.1	(5)	73.4	(7)	40.2	(4)	13.3	(8)	na		132
1994-95	106.5	(7)	71.4	(5)	47.7	(5)	27.0	(16)	0.85	na	133
1995-96	90.4	(3)	68.1	(2)	56.6	(7)	12.9	(17)	0.72	na	134
1996-97	85.3	(3)	64.0	(3)	49.4	(6)	14.8	(13)	0.72	na	128
1997-98	102.6	(1)	67.1	(2)	47.3	(5)	26.9	(5)	0.85	na	130
1998-99	92.3	(3)	63.4	(3)	44.3	(9)	21.3	(5)	0.85	na	133
1999-2000	73.5	(4)	54.1	(4)	35.2	(10)	13.1	(16)	0.98	na	130

(Source ABARE, 2002)

as per the definition in this report. f Costs to AFMA of managing the fishery. na. Not available. Figures in brackets refer to standard errors.

Strong property rights and effective resource allocation

The development of property rights was explored earlier. The allocation of the Class-A and B rights in 1984 were made directly by Government. Class-B units were allocated on historical activity in the NPF and Class-A units on vessel characteristics as discussed previously. The process for the subsequent transfer of these rights to a gear based system for the second part of the 2000 season, involved a review of the process by an independent AAP. Typically, as was the case of the NPF, AAP panels consist of a Judge as Chair, a technical adviser (usually an economist) and an Independent Fishing Industry Member. The AAP is provided with a full brief and undertakes an extensive round of industry and other stakeholder consultation. The Chair then provides independent advice to the AFMA Board on which allocation decisions are based. This process ensures transparency, provides opportunities for input by all stakeholders and reduces cost, delays and uncertainty in the Fishery as a result of legal claims against AFMA. Exceptionally, AFMA Board decisions based on AAP advice have been successfully challenged in the courts, requiring changes to the allocation and leading to instability and uncertainty within industry.

A major strengthening of NPF property rights occurred in 1995 when under a new fisheries management plan Class-A and units became SFRs. A comparison of the previous and new (current) property rights in the Fishery are provided in Table 3 below.

Table 3: Changes in the nature of property rights in the NPF with the introduction of SFRs

Issue	Pre-1995 NPF Management Plan	1995 NPF Management Plan
Quantum of right	Various Class-A and B holdings by individual and fleet operators	No change between plans
Period of issue	12 months	For the life of the management plan
Approval for renewal	At Minister's discretion	Automatically rolled over with new/revised management plans, i.e. no need for renewal
Compensation for removal of right	No provision	Considered on a case by case basis and would be subject to the outcome of appeal

Joint ownership of problems and solutions

As the 2001-2006 NPF Strategic Plan states: 'The cooperation of industry, managers, conservationists and researchers involved in the NPF has provided the framework to pursue ecological and economic sustainability in the Fishery.' There are a number of examples that illustrate this engagement:
- completion of a NORMAC-initiated strategic plan, with performance measures;
- development of an environmental management plan dealing with ecological sustainability issues including bycatch and marine protected areas;
- an industry initiated (through NORMAC and AFMA) ban on the retention of and products from sharks, rays and sawfishes in recognition of the growing global concern over the conservation status and the effects of fishing on these species;
- the introduction of TEDs and BRDs; and
- recent cooperation between industry members and researchers conducting fishery independent surveys.

Not all the management problems and their solutions have been accepted or embraced with enthusiasm by industry, including the introduction of TEDs and BRDs. Initially they were strongly resisted on the basis that the estimated 5% prawn losses expected to accompany their introduction would severely reduce profitability. AFMA undertook a lengthy consultation and discussion process with industry based on extensive cooperative TED/BRD research and development programmes including sea trials and at-sea extension activities. As a result of this consultative, hands-on approach, industry became persuaded of the value and benefits to their own operations of TEDs and BRDs (which are now compulsory) and found them to be generally positive for their overall operation[37].

---

37 On the grounds that the advantages resulting from higher quality (and more valuable prawns) and safer operations with the virtual elimination of having to handle large rays, sharks etc on board outweigh the costs.

Cost effective monitoring, control and surveillance

AFMA utilises a Compliance Plan for the NPF to assess the potential risks of non-compliance with management arrangements in the Fishery and provides compliance programs to counter identified risk. The Compliance Plan is reviewed and input provided by industry and the MAC before it is finalised by AFMA. The focus of the Plan is to encourage operators to comply with management requirements and catch limits, and allow for monitoring of protected species. The annual information handbook provided both in hard copy and compact disc format is useful for ensuring that operators are aware of regulations within the NPF. Pre-season briefings are provided to skippers both as a method of communication as well as a means of trying to encourage the skippers to comply with regulations – bearing in mind that skippers and the owners of SFRs are generally two different sets of people. A programme of at-sea inspections and in-port measuring (of nets) is in place under the Compliance Plan.

In 1998, a compulsory satellite-based vessel monitoring system (VMS[38]) was introduced. The system has reduced compliance costs and allowed for more efficient means of opening the Fishery[39], enforcing closed areas and targeting compliance activity. It has also improved the flow of information to the fleet from AFMA and provided details to researchers and managers concerning the fine-scale distribution of effort in the Fishery. While initially resisted by some operators, the VMS is now well accepted, with company vessel managers now using it to monitor the activity of their fleet.

Cost recovery of management costs and contribution to research

As discussed earlier, Australia's Commonwealth fisheries are managed to fully recover the attributable costs of management. NORMAC plays an important role in the preparation and review of annual budgets for the Fishery, participating in an open and transparent process which sees detailed cost information made available to all stakeholders. Industry also makes a significant contribution to research through levies.

In achieving a substantial level of contribution, the concomitant involvement of industry as a partner (with other stakeholders) in the management process has added a level of scrutiny to the budgetary process that has arguably increased the value for money from management measures. This is particularly evident in identifying priority areas for funding rather than accepting expenditure according to existing management structures.

---

38 The VMS is a tracking device using a vessel based beacon and satellite-based technology to transmit data. Vessels in the NPF are remotely 'polled' by AFMA and their positions automatically sent to a computer hubsite in Canberra.

39 Formerly, the NPF fleet would assemble off Weipa, Karumba, Gove and Darwin and at a given signal (the firing of a flare) would set off looking for banana prawn boils. This was known as the 'Le Mans' start. VMS now allows vessels to gather in a number of designated (and closely monitored) areas and the start can be monitored remotely in Canberra at considerably lower cost.

Key management features supporting management success in the NPF

This Section considers a number of key features of the management of the NPF that have underpinned the successes outlined previously.

NORMAC and the AFMA partnership approach

The AFMA partnership approach, NORMAC and its associated committees, and a comprehensive consultation and extension network have resulted in the effective engagement of all stakeholders in the management of the NPF. This has been assisted by a relatively low turnover of senior NPF fisheries management staff at AFMA and of scientists at CSIRO, resulting in the development of a strong body of corporate knowledge of the Fishery and its management. Similarly, from the industry side, many NORMAC members have been on the MAC for in excess of ten years.

Central to the NPF/AFMA partnership approach has been NORMAC. NORMAC, whose meetings are open to observers, has acted as the main point for discussion of key matters relating to the management of the Fishery. NORMAC has had to deal with difficult issues including uncertainties over stock assessment, biological and economic overfishing of stocks, methods of effort reduction, and increasing scrutiny of broader environmental impacts of the Fishery.

The discussions at NORMAC become very intense at times, due primarily to differences in opinion between the Queensland-based fleet, who generally have smaller operations, (some of which remain owner-operated) and the larger fleet-based operations associated with companies in Western Australian. These differences have in the past delayed reductions in effort proposed by NPFAG, and led to the recent Senate Inquiry discussed above. The strength of NORMAC is that it allows for the full and frank discussion of stock and management issues and an opportunity for input from all stakeholders. The AFMA Board takes the views of NORMAC, which include minority positions where a unanimous recommendation can not be reached, when making management decisions for the Fishery.

AFMA and NORMAC as key 'partners' manage the NPF with little or no political intervention. In the ten years of AFMA's operation the Minister has not had to directly intervene by issuing a Ministerial direction. This illustrates a high degree of confidence by the Australian Government in the ability of the AFMA Board and staff, in cooperation with stakeholders, to develop and implement effective fisheries management arrangements for the NPF.

In contrast with a number of other fisheries in Australia and globally, relationships between environmental interests and the NPF industry are generally good. The Research and Environment Committee of NORMAC (see Figure 3) has provided a 'think tank' to focus discussion on conservation issues, resulting in the adoption of practical and pragmatic means of dealing with such issues, while maintaining a fishing industry that is sufficiently profitable to adopt necessary measures. Increasingly, an ecosystem-based approach to NPF management and research is being adopted through collaboration with other non-fisheries stakeholders. An example of this is the action taken to prevent pollution and damage to nursery beds by ports and other coastal development, particularly in the Gulf of Carpentaria area.

The requirement that all AFMA fisheries management plans must go out for a statutory period of public consultation, and that comments received must be considered, allows for a broad range of inputs from the community on the management of the Fishery, including environmental issues.

Strong property rights

Statutory fishing rights in the NPF is the key driver for industry taking a long-term view of the Fishery. Industry, unquestionably driven by a vested interest to secure, or prevent the erosion of, the value their property rights, have generally opted to address issues in a proactive manner.

SFRs have provided a stable investment climate in which industry has invested considerable funds in reducing fishing capacity, commissioned research and dealt with environmental issues, including the implementation of bycatch reduction strategies and a programme of fishery independent monitoring including at-sea surveys.

The NPF experience has demonstrated that property rights do not always result in a common view among industry for future management of the Fishery. There are those who wish to develop, those who want to invest and others who do not wish to change their existing operations. Strong property rights in the NPF have focused attention on being proactive and moving forward while at the same time allowing scope for others who choose to lease, sell, or not change their operations.

Accountability, communication and transparency

Responsibility for the performance of AFMA rests with its Board of Directors. In terms of accountability AFMA is required under legislation[40] to:
- develop a corporate plan by 1 May each year;
- develop an operational plan by 1 June each year;
- prepare annual reports and present them to the peak industry body; and
- hold public meetings, at least on an annual basis.

AFMA's performance against targets is published in its Annual Report, down to fishery level, including the NPF.

NORMAC's Strategic Plan also includes performance measures, which are reported upon by AFMA on a six-monthly basis. While there are no specified penalties for non-performance, considerable criticism and political fall-out is likely if there are significant shortfalls in performance.

An open and transparent process within NORMAC provides access by all stakeholders and the wider community to the deliberations of the Committee. Prior to each NORMAC meeting, an industry association meeting is held, ensuring that

---

40 The Fisheries Administration Act 1991.

industry members of NORMAC are full conversant with the views of their peers[41]. NORMAC papers and records are placed on the AFMA website and all participants in the Fishery are provided with regular reports and data.

Due to budget constraints there is now no longer a full-time CSIRO officer or AFMA logbook coordinators working with industry, and in recent years there has been a lower level of activity in, and requirement for fieldwork, particularly in the area of prawn biology. Funds have also been constraining on the collection of fisheries independent data from research cruises. This has tended to reduce the profile of CSIRO and fisheries managers in the field as attention has turned to stock assessment and the environmental or whole-of-fishery issues associated with trawling in the NPF. The newly implemented 50% industry funded programme of fishery independent surveys commenced in 2002 will increase the field presence of CSIRO scientists and provide valuable data to augment stock assessments, particularly at a species level. These surveys will eventually be fully industry funded.

In terms of research and data collection, considerable effort has gone into working with fishers to gain co-operation for, and support of, research activities, and ownership of the results of such research. For example, the introduction of TEDs and BRDs was assisted by the substantial involvement of gear technologists from research agencies. These technologists provided on-board technical assistance to industry to reduce prawn losses and make these devices work more efficiently. This built a better understanding of the process of bycatch reduction and was pivotal to the eventual successful introduction of TEDs and BRDs to the whole NPF fleet.

Also central to good industry relationships has been the use of industry pre-season workshops at which an effective[42] method of allowing all fishers to have their say has been adopted.

Managing to a road map; legislative objectives[43] and management planning

The way ahead for the Fishery is well laid out both in the NPF Statutory Management Plan and the complementary NORMAC NPF Strategic Plan. Through the consultative processes outlined above, NPF stakeholders have agreed to long-term management objectives for the Fishery, consistent AFMA's overall legislative objec-

---

41 NORMAC members are chosen on an expertise rather than representational basis. That said, the NORMAC membership has traditionally had one member from each of the four key industry organisations.
42 Attendance at the workshops often exceeds 100, creating an intimidating environment for some Industry participants to make points on research results or associated management responses. To address this, participants are broken up into small tables with a group of around ten individuals to a table. Each group is chosen carefully to represent different Industry sectors, scientists and other stakeholders. After brief research and other presentations, each table is given key issues to discuss and formulate positions on, which are presented by the elected spokesperson for the table. This allows all individuals to have input to the process in a low-key manner, and without having to take the floor to do so.
43 Summarised from the Fisheries Management Act 1991 and the Fisheries Administration Act 1991.

tives. AFMA and NORMAC are bound to a comprehensive review process to measure performance against those objectives. While these plans exist, there is still occasional industry frustration at the frequent changes in management measures that have characterised the adaptive management of the NPF.

Research planning in the NPF has also been comprehensive and has generally ensured that research projects are generated and funded to underpin management priorities. An excellent example exists in the results of a recent bycatch study conducted by CSIRO in cooperation with other researchers. The study found that of the 411 species taken as bycatch, 13 were assessed to be a 'high priority for management, monitoring and research' in terms of their being the least likely to be sustainable (Stobudzki et al, 2000). Attention is now being focused on finding out more about those species, rather than attempting to research and 'manage' the whole suite of bycatch species.

A committed and mature industry

The NPF is the Australian Commonwealth's most valuable (and arguably the most consistently profitable) fishery which makes it more worthwhile and important for industry to be proactive in addressing challenges. Strong industry leaders and companies are committed to best environmental practice and innovation as a necessary complement to their long-term financial interest in the Fishery. A number of the leading companies in the NPF have diversified interests in the seafood sector and this diversification may help them absorb gear reductions and other operating restrictions over the short term. In addition, they have the ability to mount a strong political lobby and articulate positions within the complex bureaucracy and legislation that governs the Fishery.

Where differences exist as to the extent of stock problems and how best to deal with them, the NPF industry has been able to reach consensus on management action and has advised the Board to take the necessary actions required to maintain sustainability and profitability in the Fishery.

Contrary to many other fisheries, the NPF has tended to accept (to a greater or lesser extent) fisheries management problems and then spend time working on strategic solutions, rather than fighting the inevitable. Much of this longer-term view has arisen from strong property rights and a strongly vested interest in protecting future long-term asset values and income streams.

Good fisheries data

Catch and effort data have played a vital role in underpinning research and informing management decisions within the NPF. Although costly, face to face contact with fishers was pivotal to good logbook data, especially through the work of logbook officers in the 1980s and early 1990s.

Data have been collected in two main areas; from vessels via logbooks and landing returns from owners and processors. Logbooks have undergone an adaptive process, reflecting the development of the Fishery towards tiger prawns in the 1970s, and since 1977 their completion became compulsory. The evolution of logbook design has involved continuous industry involvement and input, since the accuracy of the data is a function of industry satisfaction and commitment to the process. In order to secure and maintain that commitment, seemingly minor issues have been considered, including logbook size (to fit on narrow wheelhouse tables) and left and right handed opening logbooks (to allow for left-handers). Careful design and feedback to industry in the form of data summaries continues to ensure that data needs for the Fishery are designed to obtain maximum information for minimum cost. The latter is keenly monitored by industry since, under AFMA's cost-recovery policies, they pay the bill.

Data verification is now carried out through comparing logbooks with seasonal and landing (and transshipment) reports, comparison of individual logbook data with typical catches in given areas, and boarding and inspection.

A participatory and planned process for the provision of research

The NPF has a long history of participatory science, with CSIRO and other providers working closely with industry and fishery managers from the start of the Fishery. Research has generally focused on providing information on which to base management decisions, primarily in the area of designing input controls that optimise returns from the resource within an ecologically sustainable framework. Industry actively promotes the provision of accurate logbook and other fisheries information as discussed above.

The NPF stock assessment process is complex and the model and underlying assumptions have been developed (and changed) over the years, with additional tools and techniques utilised to refine estimates. Industry has been keen to receive finite 'numbers' in terms of optimal effort and catch and this requirement was generally accommodated by scientists. More recently, ranges have been provided and elements of uncertainty introduced. These concepts have not always been well received by fishers but the reality (and fallibility) of the stock assessment is now gaining wider appreciation.

Since the inception of research in the NPF, effort has been made to communicate research results to industry and other stakeholders in plain English. This communication is increased through the use of occasional pre-season workshops, usually called around particularly significant issues that need to be addressed by industry, such as capacity reduction. In such cases, the workshop provides an opportunity for full and frank debate on NPF research, its relevance and the likely impact on the future management arrangements for the Fishery. The engagement of industry in the NPF stock assessment process has been essential in winning general support, albeit not unanimous at times, for reductions in effort and other fisheries regulations including seasonal and area closures.

As research funding has shifted away from the community via government towards 'user pays' industry has increasingly footed a significant proportion of the fisheries research bill. As a major contributor to research, providing around 20% of total research funding for the Fishery, industry takes a keen interest in obtaining value for money from its research dollar.

Considerable attention is paid by NORMAC to prioritising research. NORMAC has a set of Research Priority Areas that are reviewed annually and provide a guide to researchers to the needs[44] of the Fishery. A sub-committee of NORMAC annually reviews proposals and assesses them against NORMAC's priority areas. The results of this review are provided to research funding agencies to assist with assessing the relevance of and need for the proposed research.

For longer term planning and to establish a research framework, NORMAC's five-year research plan provides a listing of research priorities, linked to management issues facing the Fishery. This plan provides guidance both to researchers seeking funding support and research funding agencies.

Finally, AFMA has used the services of outside stock assessment specialists to review the work of CSIRO and provide independent evaluation of the methodology, results and, to some extent, the management responses.

Third party assessment

To be able to export after December 1 2003, all Australian fisheries must qualify as either exempt from the Australian Wildlife Trade Operation (WTO) permit system or be declared as an approved WTO operation under permit while conditions are being met to allow it to move to exemption. To achieve exemption the NPF is undergoing third party 'strategic assessment' by the Environment Australia, a Government department[45].

The assessment criteria to achieve exemption are extensive and are based on key principles relating to the maintenance of stocks to avoid overfishing, rebuilding overfished stocks, and minimising impacts on the structure, productivity, function and biological diversity of the ecosystem. A Strategic Assessment document for the NPF has been prepared by AFMA and is currently (June 2003), under assessment by Environment Australia. The document will be used as a basis to determine if the NPF is operating under the principles of ecological sustainability, by ensuring that data collection, assessment and management responses are in place for target and bycatch species and the broader environment.

The process has encouraged a rigorous and comprehensive whole-of-fishery approach to environmental sustainability, many facets of which were already being addressed by NPF management prior to the establishment of the Strategic Assessment process.

While this process has been seen by some industry members as a threat, once exemption is achieved exported NPF prawns will become 'certified' along the lines currently being promoted by the Marine Stewardship Council. Given the interest

---

44 Given current constraints on fisheries research funds in Australia, research projects that do not have the full support of NORMAC have minimal prospects of being funded.
45 Strategic assessment is required for all Commonwealth fisheries. Those fisheries with an export component need to be assessed by Environment Australia prior to December 2003 all others before December 2005. State fisheries that have an export component will need to be strategically assessed by December 2003.

in 'green' production of food products, this could have substantial downstream economic benefits for industry.

THE FUTURE MANAGEMENT CHALLENGES FOR THE NPF

Dealing with input controls

Despite the general economic and other problems levelled against input controls, all Australian prawn fisheries remain under this form of management. Fishers will continually strive to increase efficiency and it is highly likely, if not inevitable, that further effort reductions will be required in the NPF.

Management using input controls, particularly those regulating gear and vessel inputs, requires due consideration of the technical (engineering) component of the problem of effort creep. Researchers in this field consider that in the past too little attention has been given to this component, with past advice to managers and industry being centred on the views of biologists, statisticians and economists.

The introduction of gear based Class-A SFRs is considered to provide a more flexible means of controlling effort[46]. They are, in effect individual transferable effort units (ITEs) and it is claimed that operators will more easily be able to adjust their operations to allow for effort creep, or decisions taken on basis of stock assessment advice.

This view was not shared by sectors of industry as evidenced by the lengthy appeal process through the AAP and the Senate Inquiry. There were some that considered that gear units are fundamentally flawed and that a control on horsepower would provide a more effective and equitable option[47]. The degree to which vessels are able to increase towing speed in the face of gear reductions, and thereby attempt to maintain catch, will be a key issue to watch in the future. While it may be easier to measure changes to effort within the Fishery, the ability of headline length controls to constrain effort and the differential economic impact of gear SFR reductions between vessel sizes will remain complex issues.

There is little doubt that arguments can be raised on whatever basis of input controls are put forward; such is the nature of this type of management regime.

There has not been a stock assessment since 2001 and the next is due to be discussed by NORMAC at the end of June 2003. This assessment is keenly awaited by industry to see if the results of the effort cuts are beginning to flow through to target prawn stocks, and in particular brown tigers (*Penaeus esculentis*).

Variable closures (in space and time), based on observable indicators of the

---

46 To reduce capacity in the previous system required changes to vessel characteristics which are not considered to be easy to achieve in a flexible and economically efficient manner, since it is easier to reduce headline length than to alter a vessel hull or engine capacity.

47 The work by David Sterling presented at the Senate Inquiry predicts that with restricted headrope lengths and net configuration (twin-rig) industry will seek to improve performance by substituting more of the key unregulated input; that is, engine size.

current status of the Fishery instead of fixed closures, have been considered. Such closures would allow the Fishery to react in real-time to, for instance, lower or higher than average stock levels by shortening and lengthening the season, respectively. Preliminary modelling of this form of closure has indicated that the economic benefits are likely to be negligible. This situation may change as the understanding of biological relationships within the Fishery improves.

The next few years will determine if the current input control regulations are effective in constraining effort and providing a means of efficient adjustment when the almost inevitable effort creep reoccurs.

Alternative management strategies

AFMA's policy document (DPIE, 1989) states that 'The Government believes that because individual transferable quotas (ITQs) are a management tool which allows autonomous adjustment of fishing fleets, these should be the preferred management tool'. The document goes on to state that where ITQs are less cost effective than other measures, then any other measures must include some mechanism for removing capacity. While input controls are problematic due primarily to the issues of input substitution, effort creep, and other inherent inefficiencies for operators, they are likely to remain AFMA's preferred management mechanism for the NPF, at least in the medium term.

The most frequently stated reasons for not moving towards the alternative of setting a total allowable catch (TAC) and allocating a proportion (quota) of that TAC to NPF operators as ITQs are:
- the need for frequent changes to the TAC to avoid overfishing in poor recruitment years and loss of production in good years because the Fishery is usually based on a singe year-class and prawns have widely fluctuating annual recruitment, particularly in the case of banana prawns;
- the costs of pre-season sampling and fast-tracked stock assessment that would be required to set a TAC;
- high grading[48], leading to dumping and non-recording of low value prawns, with a flow-on impact on stock assessment; and
- the extreme difficulty and high costs of enforcing prawn catch landings as the Fishery is in a remote area, NPF vessels transship most of their product at sea, and low quantities of prawns can have very high values.

ABARE economists have a differing view[49] and consider that while the prawn Fishery has returned reasonable profits to operators over time the current input control regime may not be providing effective protection to prawn stocks and is inefficient. They consider that the benefits of the last round of cuts to effort will, as has occurred in the past, be eroded over time as operators seek to rearrange unregulated

---

48 Most commonly, an activity carried out under quota systems to select the most valuable species and sizes from a mixed catch.
49 Views of ABARE obtained during an interview by the author in Canberra, May 2003.

inputs and increase their personal production. ABARE have indicated that current gear and other restrictions may also be constraining operations and may lead to economic inefficiencies that could be overcome through the use of output controls. They consider tiger prawns, having inherently less seasonal variation than banana prawns, may be good candidates for a TAC/ITQ approach. ABARE are currently working on a report on the benefits of switching to output controls for tiger prawns. These views are not shared by industry or AFMA fisheries managers.

Sustainability issues

There remain some concerns for the long-term sustainability of the NPF. These concerns focus on three main areas:
- the effectiveness of effort reductions in terms of the recovery of tiger prawn stocks;
- protection of at-risk bycatch species, such as sawfish (Pristidae); and
- protection of biodiversity within the NPF region.

In all three cases, the threats have been identified and industry and researchers are currently working to address them. The degree to which Environment Australia and conservation NGOs are satisfied with these steps and the results and actions arising from current research will be a key factor in the future management of the NPF.

Uncertainty remains over the status of the tiger prawn stock, and in particular brown tigers. There is however in an informal view held by most CSIRO researchers that the NPF is getting close to the level of effort that will allow tiger prawn stocks to rebuild.

The EPBC (Strategic Assessment) process discussed earlier will provide a strong incentive to achieve full sustainability within the Fishery and therefore industry is likely to continue to take a proactive approach to ecosystem-based management.

Even if biological sustainability can be achieved, some ethical issues have been raised that may need to be faced in the future. These include the acceptability to the community of the NPF as a:
- relatively 'dirty' fishery in terms of bycatch, where the prawn: bycatch ration remains at around 8:1 in the NPF (CSIRO unpublished data), despite the introduction of BRDs; and
- an energy-expensive fishery that consumes around 2.5 litres of fuel for every kg of prawns caught[50].

The former issue is being addressed as a priority by the development of more efficient BRDs and other strategies and further data collection to determine if current levels of bycatch are sustainable. Discussion of the latter issue is beyond the scope of this paper.

---

50 Information provided by Qld East Coast prawn trawler operator, D. Sterling.

Divergent fleet interests and views

To accommodate the needs of all operators NORMAC has endeavoured to reach compromise solutions. This has included the use of 'top up' gear SFRs to lessen the impact of the changeover in management 'currency' from boat characteristics (hull size and engine power) to gear units on operators with very small boats.

At times, smaller, generally Queensland-based operators, express considerable frustration at what they perceive as the 'big end of town' corporate view of the Fishery dominating debates at NORMAC. In particular they consider that large fleet operators promote advice and decisions centred on the profits of large fleet operators and their ability to absorb the impacts of capacity reduction. The smaller and somewhat less flexible small operators who maintain an attraction towards a lifestyle type of fishery feel they will be less able to absorb these effects. Given the shortened fishing season and pressures by larger operators to increase the season and reduce the number of vessels, the small vs. corporate operator issue will continue to be significant.

The ability of the NPF to continue to adapt to changing fisheries management circumstances will require that the differences between the large fleet and smaller operators continue to be resolved and do not undermine the effectiveness of NORMAC.

CONCLUSIONS

It is almost impossible to state at a given point in time that a fishery is wholly successful in terms of its management. Marine ecosystems and fisheries are of their very nature dynamic, and this is particularly the case with short-lived species such as prawns.

One of the key successes of the NPF has been an ability to continually adapt its management with a view to long term economic and biological survival. This has created an environment of stability and commitment to the future despite continual changes in key management measures. Industry, managers and other stakeholders have generally committed to collaborative discussion of solutions to problems rather than fighting lengthy battles over whether or not the problem should be addressed. By taking such pro-active action, smart early decisions have paid dividends. Profitability in the Fishery has been maintained and solutions generally found to pre-empt crisis situations in target stocks or related ecosystem components.

Management responses in the NPF have included the reduction of effort through buy-back schemes and other means, addressing environment issues, and an increasingly extensive system of closures. These responses have been supported by a relevant and cost-effective research programme as well as low management costs relative to the value of the Fishery. In turn this pro-active form of research and management has led to a growing sense of trust between the NPF industry and conservation NGOs and agencies, and the public.

Management challenges still face the NPF today and will continue to do so for the foreseeable future. All the management features underpinning the success of the Fishery must continue to be reassessed, adapted and evolved. Remedial actions have been taken to address the current decline in tiger prawn stocks as well as to address the issues of impacts on the wider environment. The next five years or so will determine if these actions have been sufficient. Based on track history, it is likely that the NPF will continue to adapt to change and will take the research and management actions necessary to assure a long-term future.

References

ABARE. 2002: Australian Fisheries Statistics 2001. ABARE, Canberra, Australia.

AFMA. 2002: Australian Fisheries Management Authority Annual Report 2001-2003. Canberra, Australia.

AFMA. 2003: Strategic Assessment Report: Northern Prawn Fishery. AFMA, Canberra, Australia.

Brown, D., D. Galeano, W. Shafron and A. Blias 2002: Monitoring the impact of new management arrangements in the northern prawn fishery. ABARE Report of the Fisheries and Aquaculture Branch, Department of Agriculture, Fisheries and Forestry. Canberra, Australia.

BRS. 2002: Fishery Status Reports 2000-2001; Resource Assessments of Australian Commonwealth Fisheries. Bureau of Rural Sciences, Canberra, Australia.

DPI. 1982: Development and Management of the Northern Prawn Fishery. Technical working Group, Northern Fisheries Committee, Australian Fisheries Council. AGPS, Canberra, Australia.

DPI. 1989: New Directions for Commonwealth Fisheries Management in the 1990s: A Government Policy Statement. AGPS, Canberra, Australia.

Galeano, D., W. Shafron and C. Levantis 2002: Australian Fisheries Surveys Report 2001. ABARE, Canberra Australia.

Jarrett, A. 2001: Initial allocation of unitisation (boat/engine units) as harvesting rights in Australia's Northern Prawn Fishery. In: Case studies on the allocation of transferable quota rights in fisheries. *FAO Fisheries Technical Paper* 411. FAO, Rome.

Meany, F. 1993: NPF Restructuring – Lessons for the Future. *Australian Fisheries*, November 1993.

NORMAC. 2001: NPF Management – A Delicate Balancing Act. NPF Strategic Plan 2001-2006. AFMA, Canberra, Australia.

NORMAC. 2001: Five year Research Plan 2001-2006. AFMA, Canberra, Australia.

NPFAG 2000. NPF Fisheries Assessment Report. A report prepared by the Northern Prawn Fishery Assessment Group (NPFAG) during and following meetings in 1999 and 2000. Cleveland, Australia.

Pownall, P. (ed.) 1994. Australia's Northern Prawn Fishery: the first 25 years, NPF25, Cleveland, Australia.

Abbreviations

AAP	Allocation advisory panel
ABARE	Australian Bureau of Agricultural and Resource Economics
AFMA	Australian Fisheries Management Authority
AFZ	Australian Fishing Zone
BRD	Bycatch reduction device
BRS	Bureau of Rural Sciences
CPUE	Catch per unit effort
CSIRO	Commonwealth Scientific and Industrial Research Organisation
ESD	Ecologically sustainable development
EPBC	Environmental Protection and Biodiversity Conservation Act 1999
FAG	Fisheries assessment group
GPS	Global positioning system
ITQ	Individual transferable quota
MAC	Management advisory committee
NGO	Non-government organisation
NORMAC	Northern Prawn Fishery Management Advisory Committee
NPF	Northern Prawn Fishery
NPFAG	Northern Prawn Fishery Assessment Group
NPF (Qld) TA	Queensland NPF Trawler Owners' Association
SRRATLC	Senate Rural and Regional Affairs and Transport Legislation Committee
SFR	Statutory fishing right
TAC	Total allowable catch
TED	Turtle excluder device
WTO	Wildlife Trade Operation

Conclusions: Factors for Success in Fisheries Management

Stephen Cunningham[1]

Introduction

Understandably, there continues to be much pessimism concerning fisheries management world-wide. A recent letter to Nature (Myers and Worm, 2003) and a related statement by the Pew Fellows in Marine Conservation (Anon, 2003) drew attention once again to the problem of overexploitation. The headline of the associated press release (Ewire, 2003) was "Leading Experts Confirm Immediate International Action is Needed to Save World's Fisheries".

The Pew Fellows Fisheries Statement calls for action along four lines:
- Engaging institutions and stakeholders in policy-making and fisheries management,
- Managing and evaluating fisheries to promote sustainable fishing,
- Protecting and sustaining ecosystems in which fish live, and
- Preventing further ecological harm by careful use of technology.

Whilst it would be foolish to pretend that there are not problems in world fisheries, the case studies presented here demonstrate that there is reason for optimism. The studies demonstrate that in various places action is being taken, often along the lines suggested by the Pew Fellows. The results are encouraging. The action that appears most needed is to support successful initiatives and draw on them for the design of management systems elsewhere.

Whether action is necessarily needed at an international level is a moot point. Numerous international agreements and action plans now exist and the main requirement is how to put them into operation. The case studies demonstrate that there is no single recipe for success. Management arrangements must be tailored to the particular circumstances of the fishery and the country concerned. An important requirement is to ensure that the scale of the management system is commensurate with the scale of the resource, which may or may not require international action.

Given, then, the need to improve the way fisheries are managed and the information in the previous chapters, is it possible to identify the factors for success? The literature review highlighted a variety of factors based upon theoretical and empirical evidence, and these factors were borne out by the case studies.

---

1   IDDRA.

Factors for success

Creating appropriate incentives.
In unmanaged and poorly managed fisheries, incentives are created which push the fishery towards overexploitation. An important challenge for fishery management authorities is to devise management systems that remove such perverse incentives and instead encourage rational exploitation.

One approach is to develop use right systems. A range of choices is available for such systems, and the best choice will depend on the particular circumstances of the fishery. Defining use rights may help to achieve success in a number of ways. An important element is that, if well-designed, such systems give users a stake in the future of the resource which can help to improve exploitation patterns and compliance with regulations. They can also eliminate the race for fish.

For use rights to contribute to success they have to be both equitable in their allocation and seen to be a fair means of controlling effort by the wider group of stakeholders (the nation-state).

Rights can come in a wide variety of forms. The cases of the Pacific Halibut fishery, the development of Namibia's fisheries post-independence and the allocation of fishing rights in Shetland all show how formal use rights, allocated at individual or vessel level, can help reduce the race to fish. Defining fishing opportunities enables fishers to match their fishing activity to the market rather than being driven in the race to fish. In the case of Namibia, allocation of use rights in the form of quota also help build 'identity' with the new fishery which prior to independence had been a de facto open access fishery. But rights-based management does not have to manifest itself in the form of formal quotas. The study of fisheries in India and Senegal show how community norms and processes may be able to control effort and catch through assigning 'rights'. In the case of Senegal the catch levels were set by the community themselves, rather than by an outside agency.

Institutional Capacity
There is little doubt that institutions are critical to fisheries management. Biology was the original basis for management, economics and the behaviour of fishers was then added to the biological framework and, in the last decade or so, institutional approaches to understanding fisheries have come to the fore. An institutional approach to fisheries management emphasises that besides the biology and the economics, fisheries managers need to understand how a wide range of institutional arrangements impact upon the fishery. These institutional arrangements include legislative frameworks, policy processes, mechanisms for cooperation, institutions for research and information collection and analysis and so forth. But, more than just institutions, fisheries require sufficient institutional capacity to carry out these tasks. This capacity is often lacking in many developing countries, but, even so, success stories are in evidence.

The study of Mauritania demonstrates how an appropriate institutional framework was able to ensure that a significant proportion of Central Government revenue came from the fisheries sector through the collection of resource rent. Likewise the Indian case study shows how informal institutions such as the Caste Panchayats contributed towards the cooperative management of fisheries resources. Institutions are dynamic and the case of Pacific Halibut shows how institutional change (which is often very slow) was partly responsible for the long time it took fisheries managers to regulate the fishery in a manner which effectively lengthened the season and reduced risk.

Holistic approaches to fisheries management planning and stakeholder participation

Fisheries managers cannot act in isolation. Fisheries may be just one part of livelihood strategies and from an ecosystem perspective they are just one part of a bigger picture. Successful management must be cognisant of this. Where appropriate, multi-sectoral approaches must be taken so that fisheries management recognises the interests and impact of related sectors. Generally speaking the risk is that other sectors will adversely impact the fishery sector (e.g. through habitat destruction) but there is a need to take account of possible negative impacts by the fishery sector, again to avoid challenges to the management system. As part of this holistic approach, stakeholder participation is critical.

The case study from Shetland provides an excellent example of how a fishery can be turned around when all concerned work together. Through cooperation by environmentalist, fishers and the state, significant results were achieved. In India, the caste panchayats are charged not with managing the fishery, but with managing the community – taking account of all the constituent parts. The cooperation of stakeholders is also evident in Senegal where fishers were able to work together to improve conditions on the beach and the state was willing to work with the fishers also.

Dealing with complexity and change in fisheries management

Fisheries are located within diverse and complex systems. Biology, sociology, economics and institutions all bring pressure to bear on the system. Complexity is often charged with being the cause of failure in management systems where in fact it should be the catalyst for innovation. To successfully manage such complexity requires flexibility, the ability to learn and adapt.

The caste panchayats in India are an excellent example of how flexible such a system can be. Over a long period of time, these institutions have managed the fishery and the community that depends on it, by being able to adapt to external changes and respond to internal demands. The Namibian case-study demonstrates that through a period of considerable turmoil at independence, the fisheries management system was able to define the boundaries to the fishery (the EEZ), establish a quota system and create a sense of 'ownership' amidst a complex political and economic environment. What is more both the Namibian case and the Pacific Halibut case show how the biological complexity of a fishery can be successfully managed and a fishery brought back from a state of over-exploitation.

Cooperation in fisheries management

A fishery that is able to draw on the cooperation of all concerned will be more likely to demonstrate success than one where there is disunity. Cooperation can be both horizontal (between fishers) and vertical (between fishers, industry and the state). Cooperation can also manifest itself through co-management arrangements.

The Senegalese case study is an excellent example of how cooperation amongst fishers was able to halt stock decline, improve livelihoods, and, eventually, co-opt the merchants too. Cooperation amongst the fishers in the Northern Prawn Fishery in Australia was able to help a new management regime be put in place, something also demonstrated by the Halibut study. In Namibia, cooperation between industry and the state was instrumental in the early success of the new fisheries regime and cooperation between different stakeholder groups was able to work towards a successful resolution to the sand-eel problem in the Shetlands.

Resource rent as a central concept in fisheries management

The ability to extract resource rent and allocate it within the economy is a critical contributor to success in fisheries management for two reasons. Firstly, because if resource rent is being collected then there is a chance that the fishery is being managed effectively from an economic perspective in which case it is also likely that exploitation levels are biologically sustainable. Secondly, a fishery that is able to contribute to the wider economy through the collection of resource rent is potentially fulfilling the social objectives of a fishery management plan. Although it is often assumed that resource rent is being dissipated in the fishery, there are some notable examples of where resource rent has been collected successfully.

Mauritania has consistently financed over 20% of its Central Government expenditure with revenues from fishing. To begin with, the domestic fishing industry provided the lion's share of this revenue through the extraction of resource rents via export taxes. Since 1995, however, the situation has gradually changed and the contribution now comes almost exclusively from the fishing access agreement with the European Union. Namibia was also successful in generating resource rent from its fisheries and for most of the period since independence has succeeded in covering costs.

Policy frameworks

Management of natural resources has to sit within a national policy framework. Such a policy framework is made up of the stated objectives of the various state departments (fisheries, trade, environment etc) and the overall macro-economic goals of the government. The strength, flexibility and appropriateness of the framework will have a sizeable impact on the success of the management objectives. Policy frameworks for natural resources can range from the highly optimistic and lacking in any identifiable goals to the more realistic that identify clear goals and how and when they may be reached.

The Northern Prawn Fishery in Australia is a good example of how a good policy framework was able to set the boundaries for the fisheries management system – and how this was ably supported by a good legislative framework too.

Conclusion

The factors identified above are among the most important contributing to success. They are certainly not the only ones. Given the multi-dimensional nature of success (and failure), other factors may be more important in some places, at some times and to some people. Among other factors, one could also mention a perception of fairness amongst stakeholders. This may be engendered through cooperation and co-management arrangements as mentioned above. The development of a process of fishery management planning is also important – a process which needs to be on-going and reflective of its context. Enforcement, compliance, monitoring and surveillance are also all elements that are needed in a successful fishery if the policy is to be efficiently put into place, if the legislative framework is to be respected and if biological targets are to be met.

In addition to being multi-dimensional, success is also dynamic. It is never achieved definitively. Fishing takes place in a very dynamic environment, both in a physical and in an economic sense. A successful management system is one that can cope with the dynamics, either through its inherent design or through adaptation of management measures or by some combination of the two. The management features underpinning success must be continually reassessed, adapted and evolved.

This study has generated a great number of lessons for successful management. Perhaps the most important is that there is no simple, unique recipe for success. At the same time, it is important to recognise that fisheries exploitation is everywhere an economic activity, undertaken to achieve social objectives. Such objectives have two broad forms. First, the fishers themselves attempt to achieve their private social objectives through their exploitation of fish resources. Second, the management authorities attempt to achieve public social objectives through the way in which they decide to manage the activity. The achievement of private and public social objectives depends on the economic success of the activity. This economic success in turn depends on the institutional framework within which the activity takes place and on the sustainability of the resource base and its environment. The evidence suggests that success cannot be achieved by focussing on any one of this dimensions. Success in fisheries management comes when all of these elements are explicitly (or sometimes implicitly) recognised within an overall management framework.

REFERENCES

Anon. 2003. Pew Marine Conservation Fellows' Action Statement for Fisheries Conservation http://www.pewmarine.org/pdf/Fisheries_Statement.PDF

ewire 2003. Press Release http://www.ewire.com/display.cfm/Wire_ID/1610

Myers, R. and B. Worm. 2003. Rapid worldwide depletion of predatory fish communities. Nature 423: 280-283.